Non-Intrusive Methodologies for Large Area Urban Research

Non-Intrusive Methodologies for Large Area Urban Research

I. P. Haynes, T. Ravasi, S. Kay,
S. Piro, and P. Liverani

Archaeopress Archaeology

Archaeopress Publishing Ltd
Summertown Pavilion
18-24 Middle Way
Summertown
Oxford OX2 7LG

www.archaeopress.com

ISBN 978-1-80327-446-1
ISBN 978-1-80327-447-8 (e-Pdf)

Cover: Images of the Rome Transformed Team in action, courtesy of members of the Rome Transformed Project.

This book is available direct from Archaeopress or from our website www.archaeopress.com

This volume is dedicated to the memory of our
friend and esteemed colleague,
Daniela Zamuner.

Rome Transformed has received funding from the European Research Council (ERC) under H2020-EU.1.1., the European Union's Horizon 2020 research and innovation programme (Grant agreement No.: 835271)

Contents

Introduction

I. P. Haynes

At the time of writing, 4.4 billion people (56% of the world's population) live in cities and, according to the World Bank, as many as seven in ten will do so by 2050.[1] Cities, central to so many lives and debates, look set to play an increasingly dominant role.

While the global challenges associated are now rightly attracting ever more concern, scholarly interest in cities is hardly new. The defining role of cities in driving the development of civilisations, creating new forms of society, and shaping communication and exchange have long been widely discussed. A growing body of research into pre-modern cities sets the trajectories of contemporary urbanism in broader context and illuminates long-term trends in powerful ways. Just as cities are growing exponentially, so too are the ways in which they can be studied.

This volume brings together contributions by scholars committed to illuminating ancient and medieval cities through the application of non-intrusive methods. Several of the exciting projects here speak to the extraordinary results that these methods can yield at sites which were partially or entirely abandoned. Here the capacity of remote sensing systems to visualise topography, and sometimes, to capture the pulse of urban development, can create remarkably vivid insights. Such work is driving archaeological approaches to cities in an exceptionally powerful way.

In an increasingly urbanised environment, however, the capacity of cities to veil and destroy evidence for their own creation and evolution continues to introduce challenges. In temporal terms, the most successful cities are often the ones that are most problematic. Longevity is frequently characterised by radical changes in ground surface and the widespread truncation or obliteration of vital evidence. Archaeological excavations that pursue evidence for the stories of cities have developed methods to manage work at depth within intercutting multi-period deposits.[2] Yet for all the information that it does bring, excavation is expensive, and in many fast-moving urban areas, impracticable. At the same time, the quantities of other data relevant to a city's history can be remarkably diverse, and – along with the rapid pace of development – potentially overwhelm attempts to catalogue and synthesise.

Non-intrusive, and less-intrusive forms of data acquisition have a vital role to play, but in living cities their deployment must also factor in the needs of contemporary residents. Evolving and growing need for key services, for utilities, road infrastructure, and housing must all be considered. Such growth can limit the areas available for research, but it may also facilitate it. The same is true of data integration, the complex process whereby diverse sources of information can be brought together in ways that can illuminate the past and facilitate the development of the present.

The papers assembled in this volume spring from a larger conference held in association with the Rome Transformed Project. This project aims to use largely non-intrusive methods to understand the development of Rome over eight centuries (from the first to eight centuries CE). It explores 68 hectares of the city in an area that includes, alongside multiple other structures, a 1.5 km long tract of the Aurelian Walls and 0.67 kms of the Claudio-Neronian aqueduct. To achieve its goals, it integrates an extensive array of documentary sources, architectural analysis, and the investigation of 12 sub-surface excavated areas with the largest unified laser scanning and geophysical survey programme (the latter over an area of 12.5 hectares) ever conducted in Rome.

---

[1] The World Bank 2022 'Urban Development Overview'.
[2] Perhaps the best known such example is that of the Museum of London's Single Context Record system (Westman 1994).

Rome Transformed has received funding from the European Research Council (ERC) under H2020-EU.1.1., the European Union's Horizon 2020 research and innovation programme (Grant agreement No.: 835271) and as part of its mission, it emphasises the dissemination not simply of research output, but also of ideas of best practice. To that end, the Rome Transformed team facilitated a major conference in July 2021 to exchange ideas with leading practitioners about how to take non-intrusive work in urban areas forward. The papers here stem from that event, but for various reasons not all of the contributors to the conference were able to present their papers for publication.

It falls to me here to thank all those who took part in the conference, together with those colleagues who worked on the organisation of the event, notably the IT services at Newcastle University, Drs Thea Ravasi and Francesca Carboni, together with Phyllida Bailey, Elettra Santucci and Roxana Montazerian, and also those who have shared the work on the editing of this volume, Prof. Paolo Liverani, Dr Thea Ravasi, Stephen Kay and Dr Salvatore Piro. Finally, it is a pleasure to record our debt to our colleagues at Archaeopress for their generous collaboration.

References

The World Bank 2022 'Urban Development Overview' https://www.worldbank.org/en/topic/urbandevelopment/overview (last consulted 15.11.22)

Westman, A. (ed.) 1994 *Archaeological Site Manual*, MoLAS, London

RT3D stratigraphies: analysis and software design to manage data

V. Bologna, M. Azzari

LabGEO, Dept. SAGAS, Universitá degli Studi di Firenze (Italy)
Margherita Azzari - margherita.azzari@unifi.it
Vincenzo Bologna - vincenzo.bologna@unifi.it

The main goal of the LabGEO team is to build an efficient, flexible, and scalable working tool for the generation of Digital Terrain Models (DTM) across the time phases identified by the Rome Transformed project, starting from the stratigraphic data acquired through the project's survey methods.

In the absence of a software that fits the intended purpose, the team has developed a software solution, creating a tool able to manage stratigraphic data. The first generation of this tool is known as RT3D.

We started by defining an ideal project workflow, involving the teams of specialists within the project in a process of continuous brainstorming, to ensure the correct import and conversion flow for each category of data: structural analysis, geophysical survey, archival research, and environmental analysis.

Once the analysis and assessment of the available data and research needs were completed, we developed import scripts (for batch spreadsheet importing flows) and input forms to populate our database. We chose to use PostgreSQL with the PostGIS extension as the database that will be fully compliant with the GIS platforms used by the Rome Transformed project: QGIS and ESRI ArcGIS.

Specifically, the PostgreSQL software provides the following features: the archiving and normalization of the data acquired, the possibility to generate 3D geometries (with Z dimensions) that can be associated with the data, the lemmatization of the data related to interpretations (time phases), of the typologies, of the methodologies, and of the survey areas, and finally the tools to export the normalized data to the project's ArcGIS geodatabase (Figure 1).

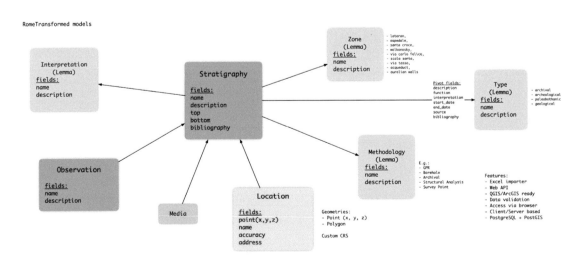

Figure 1. Database schema.

Regarding the choice of geometry and synthesis of the imported stratigraphic data, we have identified the following generation scheme: PointZ for archival data, MultiPointZ for borehole analysis and PolygonZ for data coming from the structural analysis and GPR surveys. These 3D geometries are needed to generate a DTM in a GIS environment.

We have also developed scripts to ensure a stable and automatic data import of the information coming from the project's existing datasets, to create the corresponding cards on the project database. The development of the scripts ensures that over time the import of the cards can be scheduled and that it minimizes the filing activities from one database to another. The import scripts involve loading sets based on an Excel (XLS, XLSX) or CSV structure with corresponding fields established by each team (Figure 2).

The 3D visualization tool has been designed and developed in such a way that prior to its export for use in a GIS environment, it permits easy visualisation of the distribution in space and the geospatial relationship of all the stratigraphy of the subsurface. This tool enables visualisation of all the layers identified by colour according to the typology of the data and their interpretation (Figure 3).

All the data processed by the RT3D software can be exported using two distinct exporting modes: WFS (Web Features Service) and ESRI Shapefile. From the amount of data and the typical workflow expected for these tasks, we identified Shapefile export as the most flexible data interchange tool. The tool is under development, and it will be possible to use it directly within the WebApp interface of the stratigraphy, depending on the selection of the interpretation associated with the data. In other words, all geometries generated for a homogeneous dataset (same type of geometry) for a specific historical period associated with that data (e.g.: Medieval Period - MultiPointZ) are collected. Once the Shapefile is generated, it is downloaded directly from the browser as a zip file containing the three files provided (dbf, shp and shx). These files can be imported directly into ArcGIS and contain not only the expected geometries but also the table with all the stratigraphic information.

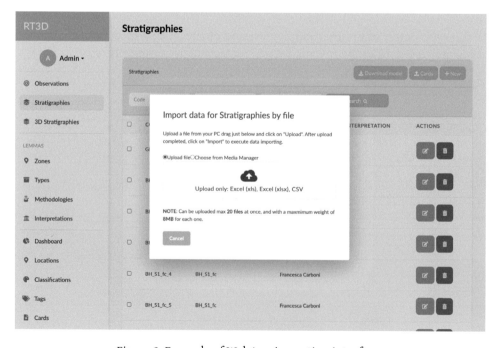

Figure 2. Example of WebApp importing interface.

Figure 3. 3D Visualizer Preview Example.

Acknowledgement

'Rome Transformed' has received funding from the European Research Council (ERC) under the European Union's Horizon 2020 research and innovation programme (grant agreement no. 835271).

Roman buildings on the western slopes of the Capitol. Investigations and new approach technologies

E. Bianchi[1], A. Pansini[2]

[1] Sovrintendenza Capitolina ai Beni Culturali, Rome (Italy)
[2] Universitá La Sapienza, Rome (Italy)
Antonella Pansini - antonella.pansini1@gmail.com

The western slopes are perhaps the least known sector of the ancient topography of the Capitoline Hill. A few years after the construction of the Monument to *Vittorio Emanuele II* (1885-1911), as part of the urban transformations desired by Mussolini, in 1926 in the entire area between Piazza Venezia and the Tiber was remodelled. Work was undertaken to isolate the Campidoglio and construct the Via del Mare, the new road artery which had to connect Rome to Ostia. The demolition works, carried out and completed in 1943, were conducted by the Antiquities and Fine Arts Office of the *X Ripartizione* of the *Governatorato* under the leadership of Antonio Muñoz.

For this purpose, an entire medieval quarter, made up of several churches and civil buildings, was demolished with the result of the rediscovery of an extensive district of the imperial age.

In 1928 *Via di Giulio Romano*, located on the southern side of the Vittoriano with all its historic buildings, disappeared and the structures of the *Insula dell'Ara Coeli*, the most representative example of multi-storey civil building preserved in Rome, were brought to light. In the Insula it was possible to recognize the origin of the modern house, as pointed out in essays by Guido Calza written in those same years.

What had been the narrow *Via Tor de' Specchi* became the first section of the sumptuous Via del Mare, inaugurated by Mussolini on 28th October 1929. After demolishing the houses, palaces and churches of Via Tor de Specchi, the tufa cliff was revealed and huge remains of residential and commercial buildings were found: the *Taberna delle Tre Pile*, the *Casa dei Molini*, the so-called *Grande Insula*, the so-called *Casa Cristiana*, the *Balneum*. Given the precarious state of conservation of the buildings, none of them was left in sight, and the decision was made to bury the remains under the new road network, now called *Via del Teatro di Marcello*.

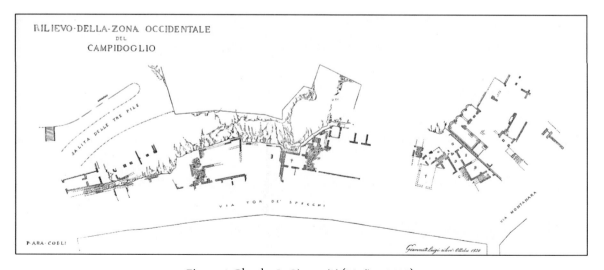

Figure 1. Plan by L. Giammiti (Muñoz 1930)

The events that have affected this area have been almost forgotten, so in 2014 the *Sovrintendenza Capitolina* set up an interdisciplinary research group made up of numerous scholars, including colleagues from Sapienza University of Rome, Southampton University, University of Roma Tre, University of Lucca and University of Calabria, as well as the INGV (National Institute of Geophysics and Volcanology). The research was aimed at examining the different historical-topographical and architectural aspects of the different phases attested in the area, as well as the finds preserved or documented in the excavations, with the aim of reconstructing the history of this district from the imperial age, through the levels of medieval and Renaissance occupation, up to the years preceding the interventions of the Fascist era.

Extensive documentation on the 'excavations of *Via Tor de' Specchi*' is now available in various archives and is mainly contained in the notes drawn-up by the official archaeologist A. M. Colini. However, it was necessary to provide for an extensive plan of general and detailed surveys and photographic documentation of the archaeological evidence still existing today. For the *Insula dell'Ara Coeli* fundamental aid was given by some university theses focused on the survey with both traditional method and with modern technologies (laser scanner), subsequently carried out systematically on all the other buildings located below Via del Teatro di Marcello. For this work, sometimes carried out many meters below the road level, it was essential to be able to count on the support of the speleologists of the Association *Roma Sotterranea* and to develop a procedure that would allow all activities to be carried out safely, according to the criteria currently required by the legislation on confined spaces.

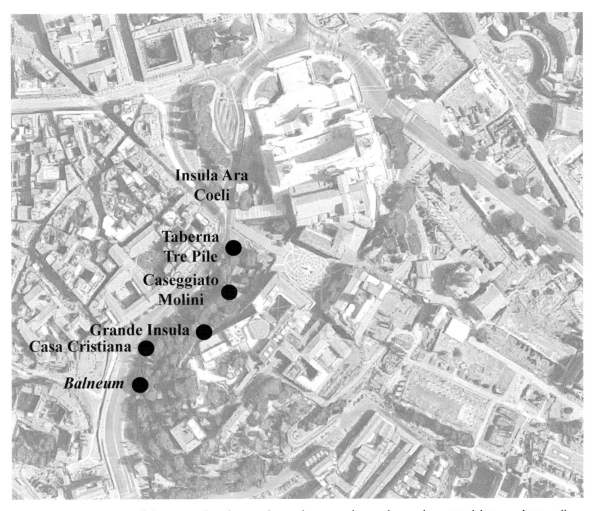

Figure 2. Positioning of the access hatches to the underground complexes along *Via del Teatro di Marcello*.

In collaboration with the University of Southampton it was possible to carry out an accurate laser scanner survey of the *Insula dell'Ara Coeli* and important geophysical investigations (conducted by the British School at Rome), capable of returning information on the nature and morphology of the land in ancient times and on the presence of ancient structures under the road surface so far never investigated.

The research group also includes other scholars, who are dealing in different ways with the post-ancient phases of this area, regarding medieval and modern elements, the vanished churches and the general topography of the Renaissance and modern quarter, as it appeared the demolition works. A forthcoming publication will give an account of this research and will be an opportunity to deepen some important issues concerning this area, such as, for example, the different sections of the walls in square work of *cappellaccio* probably referable to Rome's earliest city walls, the fragments of the Forma Urbis (some fragments that probably depict the *centum gradus*, the steep staircase mentioned by Tacitus as one of the accesses that in ancient times allowed the ascent to the Capitol) and a detailed study of the walls of the buildings of the imperial age that today it is finally possible to observe with a direct examination.

Issues relating to the management and enhancement of an important monument such as the *Insula dell'Ara Coeli*, located in a nerve point of the historic centre of Rome, require the planning of a restoration and valorisation project for the monument. A specialized thesis in Architectural Restoration with University of Roma Tre was recently developed on this topic and involved, in addition to the analysis of the structures and the survey of the state of conservation aimed at the recovery of the ancient building, a project aimed not only at the accessibility of the site but at the recovery of its symbolic, cultural and social meaning, with the prospect of sensitizing future visitors to a renewed interest in the uniqueness and identity of this historical artifact.

As part of the research project the task of the Sapienza University of Rome (Chair of Survey and Technical analysis of ancient monuments – Department of Science of Antiquities) has focus on one side the review of the old drawings and at the same time the elaboration of a new and complete graphic documentation.

Due to the speed with which the excavations were conducted in the 1930s, no complete archaeological documentation of the structures was produced. The project plan by L. Giammiti,

Figure 3. Laser Scanning survey of the urban context and of the archaeological complexes. Complete view (Survey and elaboration by G. Casazza, L. Mazzoni, A. Pansini).

published in Muñoz (1930; Figure 1), was the only drawing in which it was possible to see the planovolumetric development of the underground sites. The plan, however, lacks any other data useful to the understanding of the remains, such as, for example, levels, indications of the different type of building techniques, relationships between walls, etc.

The descriptions contained in the above-mentioned book by Muñoz, the only systematic publication of the excavations, are also very concise and often lacking this data: the text is not accompanied by elevation drawings or sections, except for a drawing by M. Barosso in which the façade of the inner courtyard of the Grande Insula is represented, and for other freehand drawings. The sketches contained in *Appunti degli Scavi di Roma*, by A. M. Colini, are very important documents: they are in fact very often full of information, reporting the thicknesses and heights of the walls, and paying attention to the characteristics of the structures. The *Insula dell'Ara Coeli*, on the other hand, was more fortunate because it was accurately described in Packer (1968) and documented by Italo Gismondi: its plans, one for each floor of the building, and the sections are still an essential tool for the study of the monument.

In order to shed light on the relationship between the ancient and the modern urban context and to provide a complete and precise modern documentation of the *Insula dell'Ara Coeli* and the underground complexes, a new survey was carried out (Figure 2). A laser scanning survey, integrated within the georeferenced topographical system of the city, was undertaken using a P20 Leica Geosystems laser scanner (Figures 3-4).

The scans were carried out all along *Via del Teatro di Marcello*, from *Piazza dell'Ara Coeli* to the beginning of *Vico Jugario*, and in every single underground complex, ensuring a correct overlap between the point clouds acquired inside and outside. In the final dense point cloud, it is therefore possible to appreciate these complexes in their planimetric and altimetric relationships.

The monuments were the subject of a photogrammetric survey: high-resolution orthophotos were extracted from the three-dimensional models, with particular attention to the decorated surfaces.

The three-dimensional documentation serves as an important 'data container' available at any time on measurements, altimetry, distances, etc.: it is an important tool for the study of monuments, even 'at a distance', considering the problems related to their difficult accessibility. Plans, sections

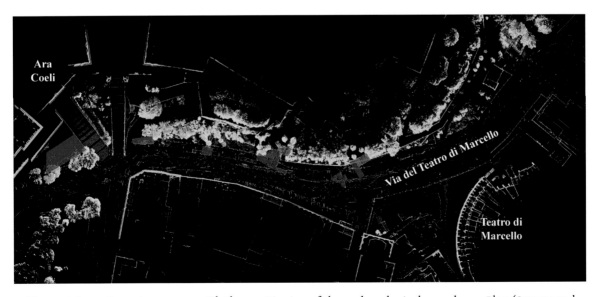

Figure 4. Laser Scanning survey with the positioning of the archaeological complexes. Plan (Survey and elaboration by G. Casazza, L. Mazzoni, A. Pansini).

and detailed elevation drawings, vectorized in CAD were extracted from it: these served as a basis for the analysis and reconstruction of the structures. Following the different types of building techniques and walls relationships it has been possible to draw up synchronic and diachronic phase plans. In many cases it was necessary to integrate the new documentation with that of the 1930s, especially for the structures no longer visible because demolished or buried. A particularly representative example is provided by the *Grande Insula* where, thanks to the overlap between the new and old survey and the documentation created during the Telecom excavations in the 90s, it was possible to reconstruct the internal and external western façade of the building.

To conclude, the realization of a new survey, carried out with the most modern techniques of documentation of the cultural heritage, combined with a critical analysis of the graphic and photographic documentation of the archive and followed by a careful analysis of the complexes, made it possible to identify the architectural peculiarities of these structures and to reconstruct the architectural and urban evolution of this sector of the ancient city of Rome.

References

Broccoli, U. 1994. *Storie di Roma tra Campidoglio e Tevere*. Rome: Telecom Italia
Colini, A. M. 1998. *Appunti degli scavi di Roma, 1. Quaderni* I bis, II bis, III, IV, Rome: Quasar
Colini, A. M. 2000. *Appunti degli scavi di Roma, 2. Quaderni V, VI, VII, VIII, IX, IX b*, Rome: Quasar
Muñoz, A. 1930. *Campidoglio*, Rome.
Packer J. E. 1968. La casa di via Giulio Romano, *Bullettino della Commissione Archeologica Comunale di Roma*, 81, 1968-1969: 127-148.

Combining past, present, and future. Non-invasive mapping for the urban archaeology of Ascoli Piceno (Italy)

F. Boschi[1], E. Giorgi[1], M. Silani[2]

[1] Università di Bologna, Dipartimento di Storia Culture Civiltà (Italy)
[2] Università della Campania Luigi Vanvitelli, Dipartimento di Lettere e Beni Culturali (Italy)
Federica Boschi - federica.boschi5@unibo.it
Enrico Giorgi - enrico.giorgi@unibo.it
Michele Silani - michele.silani@unicampania.it

The Asculum Project started in 2012 as a collaboration between the University of Bologna, the Soprintendenza Archeologia, Belle Arti e Paesaggio delle Marche and the Municipality of Ascoli Piceno, primarily as a project of urban and preventative archaeology in a long-lived city. Within this collaborative framework, the project benefits a wide range of interests, helping to reconcile the needs of preservation and research with a sustainable urban development.

The integration of new and old data from a variety of sources (geophysical and topographical survey, archaeological excavations, geological and geomorphological analysis) is at the basis of the research, which advances the reconstruction of the ancient urban landscape and its transformations over the centuries.

At Ascoli we have adopted an integrated approach for assessing complex deposits from urban contexts with continuity of life using non-destructive methodologies. The project has integrated the

Figure 1. Activities of research, education, public archaeology and preventative archaeology in collaboration with private archaeological companies within the Asculum project.

geophysical investigation of the underground deposits with the 3D documentation of the surviving historic buildings, using laser scanning and photogrammetry alongside the analysis of structural stratigraphy, the study of the building techniques and of the artefacts. Important key studies have focused on the main squares and open areas of the city, with their historical monuments (Piazza del Popolo, Piazza dell'Arengo, Piazza Viola, Piazza San Gregorio Magno), visualising remains beneath the paved surfaces and providing new insights to interpret the evolution of the buildings and the development of crucial sectors of the Roman and of the Medieval cityscape. The exploration and the assessment of the buried archaeological record is widely supported by the notable work of the archaeological units that are routinely involved in the local rescue archaeology, as a result of a consolidated relationship of mutual collaboration.

Beyond research, the project is also fostering knowledge outside the academic environment, focusing on the city and on its social and cultural development. Fulfilling the university's public engagement (or more in general towards the general call for a Public Archaeology), we have offered talks, site visits and interactive visits to archaeological excavations, to those living in Ascoli and those interested in the city's history. Some field actions were also promoted thanks to the effective involvement of cultural associations and local schools, which also took part in the planning process, sharing the objectives and promoting the required fundraising. These activities confirm the role of public engagement in raising awareness of the contribution that modern methodologies and up-to-date technologies can make in the protection of archaeological landscapes threatened by urban development, with undeniable benefits and positive impacts on the daily life of the community now and in the future (Figure 1) [E.G.].

The city of Ascoli Piceno is situated in the heart of the ancient region of *Picenum* within the valley of the river Tronto. The valley provides a natural communication route by way of the *Gole del Velino* to

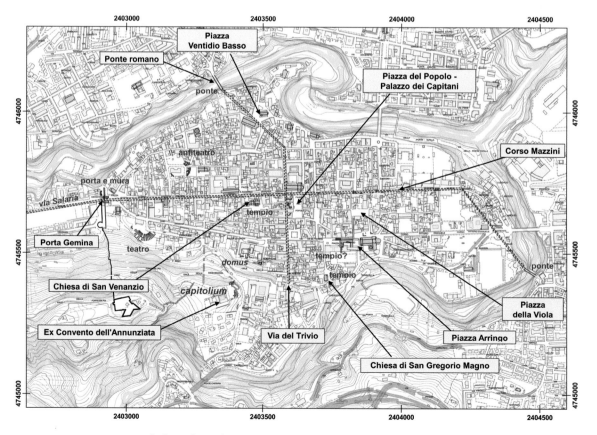

Figure 2. General plan of Ascoli with the main urban elements and ancient monuments.
Elaboration: M. Silani.

the river Tiber and thence to the Tyrrhenian sea side of the Italian peninsula. The city originated as the main settlement of the Picene culture, during the Iron Age, at the confluence of the river Tronto and its smaller tributary the River Castellano. Following its initial genesis, the city has remained in almost continuous occupation, in several places supporting standing buildings and other forms of evidence that help us to appreciate the town's evolution over the centuries. Quite often, indeed, the medieval structures, which make up part of today's urban fabric, incorporate well preserved elements of pre-existing Roman buildings, as at the church of San Gregorio Magno, lying directly above a Corinthian temple, or San Venanzio, sited above an Ionian temple. In other cases, the foundations of the medieval structures have been shown during excavation projects to directly overlie the ruined remains of Roman buildings, as at the Palazzo dei Capitani in the Piazza del Popolo or at the Palazzo dell'Arrengo in the piazza of the same name. In other situations, elements of the Roman buildings are still visible today, as at the Porta Gemina (Porta Romana); some even remain in active use to the present day, as at the Sostruzioni dell'Annunziata and the Augustan bridges of Borgo Solestà and Cecco d'Ascoli (Figure 2).

The urban complex developed at the intersection between various economic and environmental zones within the local area, including those linked to the Apennines and to the Adriatic seaboard. The connection with both the Adriatic and Tyrrhenian coasts, along with the exceptionally favourable physical geography of the settlement's chosen location, led to its precocious development as an urban focus within the surrounding region. Archaeological excavations in recent years have yielded significant physical traces of the ancient capital of the *Picentes*, well documented within the written sources for the area. This was the settlement that was initially the main ally of the Romans in resisting the Gallic advance at the transition from the 4th and 3rd centuries BC, but it later became involved in the rebellion of the *Picentes* against Roman domination in 267 BC. The remains of the allied city (*civitas foederata*) and the phase of the *municipium* are now gradually emerging as a result of important studies by the Soprintendenza Archeologia, Belle Arti e Paesaggio delle Marche. Nonetheless, the form of the Roman colony still presents many unanswered questions, such as the location of the *Forum* and of the *Capitolium*.

With the aim of answering these and other open questions the *Asculum* project adopted from its outset an integrated approach between different data sources, managed within a GIS system directly supported by 3D modelling.

Among the research promoted to advance knowledge of the ancient hidden city there are the investigations at Piazza del Popolo. Here the integration of archival studies, historical cartography, geophysical data and archaeological evidence from excavations carried out beneath the flanking Palazzo dei Capitani, led to the identification of various evolutionary phases of the main town square and to the critical rethinking of some hypotheses regarding the location of the Forum of the Roman city. The square and the adjacent palace are located near the intersection of the Via del Trivio and the Corso Mazzini, the two alignments considered by most scholars have given rise to the orthogonal layout of the Triumvirate-Augustan colony. For this reason, the area was suspected to be the possible location of the Forum square. After a phase of intense removal of building material prior to the construction of private buildings in the Middle Ages, the present square was created by demolishing the medieval structures. In the early phase, dating on documentary evidence to the second half of the 13th century, the so-called Platea Superior occupied only half of its current space. Later, in the second half of the following century, the houses and a tower in the northern section were demolished. At the same time the palaces of the nobility that lined the western side were unified into the complex of the Palazzo dei Capitani, more or less coinciding with the line of the present-day facade, while along the northern side a start was made on the great building site of the Convent of San Francesco.

This painstaking analysis, integration and depth-comparisons between the deposits uncovered in the excavations at the Palazzo dei Capitani and those recorded in the geophysical data from the

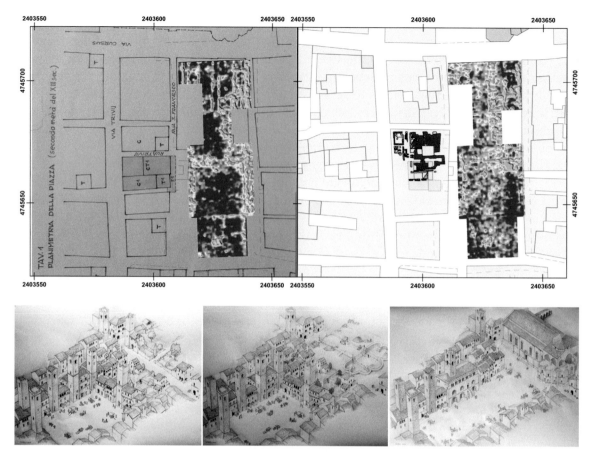

Figure 3. Overview of the integrated work in Piazza del Popolo. At the top, GPR slices from the geophysical mapping and overlap with historical cartography and the archaeological data related to the flanking Palazzo dei Capitani (Elaboration: F. Boschi). At the bottom, sequence of drawings by G. Giorgi retracing the evolution in Medieval times of the area then occupied by the square.

adjacent Piazza del Popolo, have been supplemented by a programme of laser scanning of the recently revealed archaeological evidence, the whole then being related to the GPR slice maps within a three-dimensional environment. This integration is supporting the interpretation of the geophysical data as well as assisting our understanding of the sequence of building phases, drawing on the three-dimensional and volumetric representation of the whole below-ground environment (Figure 3).

On the basis of this work, it is possible to recognize a level, about half a metre beneath the present surface of the piazza, within which there can be recognised the walls of the medieval buildings demolished to allow the creation of the square. Deeper down, at about a metre and a half beneath the present surface, a number of other features become evident and in part persist until a considerably greater depth. These deeper radar reflections can be attributed to the Roman phases attested at similar depths in the archaeological excavations nearby. Granted that these are preliminary data, still under study, it seems clear that their interpretation will be fundamental in establishing, or perhaps disproving, the existence of the Forum at this point in the urban fabric.

The presence of possible buried structures that seem to occupy a large part of the space of the future medieval square in Roman times would not seem compatible with the presence of a forum. Even with all the caution required in this preliminary phase of research, this consideration could perhaps support at least one other hypothesis. The block just to the west, in fact, would be located at the intersection of the two axes generating the Roman urban layout and would also include the

Ionic temple incorporated by the church of San Venanzio, in the present-day Piazza Bonfine. The Roman temple was located at the end of a porched square, partially preserved under the pavement of the medieval church and given its off-centre position could have been part of a larger complex that included two twin temples, perhaps at the end of an open space identifiable as the supposed forum of the Roman colony.

If so, it would be a 'foro passato', i.e., crossed by a main street, which separated the structures under the Palazzo dei Capitani (a macellum?) from the square, which we can imagine was overlooked by a basilica, and finally from the western part of the portico reserved for worship, with a scheme of functional subdivision of the various spaces typical of Roman town planning.

Moreover, we would be in the area of the town defined 'delle scaglie' in the medieval toponymy, probably due to the presence of fragments from workmen dismantling the Roman public buildings to build the new Medieval churches.

These are only hypotheses that await confirmation, which arise from archaeology without excavation, and the way in which bringing together different data opens up new research questions [F.B.].

As mentioned above, another interesting aspect of the project comes from the interaction between the private companies routinely involved in the locality's emergency archaeology and the technical staff of the Municipality directly concerned with the redevelopment or rezoning of a number of urban areas. The ethos of the project calls for a constant dialogue between these participants in the day-to-day life of the city along with research staff at the University of Bologna and the officials of the regional Soprintendenza. This kind of cooperation will hopefully protect and emphasise the need for preservation and research within the city without adversely affecting sustainable urban development.

The project in the northern part of the city is certainly a case in point. The area lies between the medieval churches of Santi Vincenzo e Anastasio and San Pietro in Castello, whose dedication and toponym highlight a settlement probably linked to the period of occupation by the Lombards. This is a part of the city overlooking the deeply sunken banks of the river Tronto, forming a loop around the promontory of San Pietro in Castello, which controlled the only way out to the other Lombard stronghold of Firmum about twenty kilometres further north. Towards the south, the same road axis continued through the city and up the Valle Castellana, where there was the well-known stronghold of Castel Trosino which, through the Salinello Gorge, led to the Teramo sector, also controlled by the Lombards. This brief analysis shows the importance of the area in the early Middle Ages, when the fragmentation between the coastal zone controlled by Ravenna and the inland one controlled by the Duchy of Spoleto favoured internal transversal routes, such as the Teramo-Ascoli-Fermo, once only a secondary branch of the Via Salaria.

The project first involved Piazza Ventidio Basso, where geophysical and archival surveys were carried out prior to stratigraphic surveys by professional archaeologists led by the Superintendency. As in Piazza del Popolo, this allowed to recognise the physiognomy of the original medieval square, called platea inferior and dedicated mainly to the cloth market. Noble towers, oratories and demolished churches were thus identified without excavation. Subsequently, stratigraphic surveys brought to light several levels of the square's pavement, which covered an early medieval necropolis, partly musealised on site. In fact, the collaboration between conservation, research and the city government led to the complete renovation of the square, which improved its urban physiognomy. The level car park has been removed and the square has been reconstructed and is once again one of the centres of gravity of city life. Finally, to complete the project, the area near San Pietro in Castello, where the car park, necessary for the residents of the historic centre, was moved, also had to be renovated.

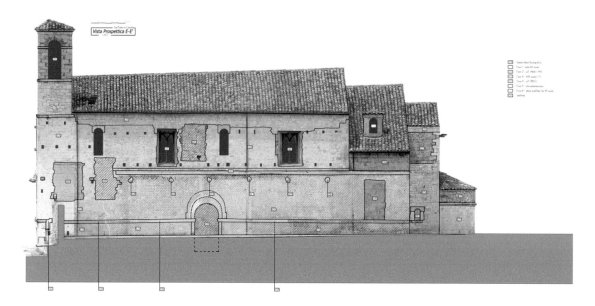

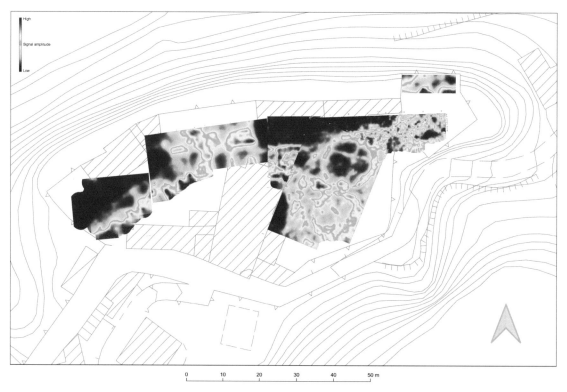

Figure 4. San Pietro in Castello and the procedure of archaeological impact assessment. At the top, archaeological analysis of the standing architectures (by F. Zoni). At the bottom, GPR survey in the whole area, inside and around the religious building (Elaboration: G. Guarino).

Among the last operations tackled with this approach there is the work led at San Pietro in Castello, launched precisely for a preventative archaeology procedure as part of the redevelopment of the area (Figure 4).

The religious building of San Pietro in Castello as a whole represents one of the various examples of late Romanesque architectural techniques of the town. However, its current appearance, and consequently its walls, is actually the result of several centuries of alterations ; today it appears as a historical palimpsest that spans the Middle Ages and the contemporary era. Within the project of reappraising the area, the archaeological potential evaluation included stratigraphic analysis of the architecture, geophysical survey and watching briefs at building sites. The combined activities underlined the presence of pre-existing structures in the façade of the church and in the postern that seem to be part of the ancient castrum, a structure certainly existing in the 11th century, as well as evidence of construction phases which are likely to be earlier than the present building's 13th century design. The results obtained seem to agree with the historical sources, which mention the first foundation of the church at the behest of Bishop Auclere in the Langobard period, along with several structural modifications over time.

Lastly, a significant contribution within the present range of investigations could come from reconstructing the natural environment at the time when the *Picentes* occupied this area before the establishment of the Roman settlement. Analytical work in the identification and interpretation of a wide range of data sources, through a combined archaeological-geological-geomorphological approach, with continuous feedback between the different data sets, is already under way with the aim of creating a topographical reconstruction of the original paleo surface that would have influenced the settlement [M.S.].

References

Bombardelli, M., Guarino, G., Massoni, M., Zoni, F. 2019. A case of rescue archaeology in Ascoli Piceno: the project of San Pietro in Castello. *Groma* 4. DOI: 10.12977/groma25

Boschi, F., Giorgi, E., Casci Ceccacci, T., Demma, F. 2020. The Urban Archaeology Project in Asculum. From civitas caput gentis to civitas foederata, in F. Boschi, E. Giorgi, F. Vermeulen (eds) *Picenum and the Ager Gallicus at the Dawn of the Roman Conquest. Landscape Archaeology and Material Culture*: 165-171. Oxford: Archaeopress Publishing Ltd.

Boschi, F. 2018. From Preventative Archaeology to the Archaeological Map. Landscape Archaeology in the heart of the Piceno. *Groma* 3: 1-14.

Boschi, F., Giorgi, E., Silani, M. 2017. Reconstructing the ancient urban landscape in a long-lived city: the project Asculum – combining research, territorial planning and preventative archaeology. *Archeologia e Calcolatori* 28: 301-309.

Giorgi, E., Demma F. 2018. Riflessioni sulla genesi e lo sviluppo urbano di Asculum nel Piceno. Dalla Città Federata alla Colonia Romana. *Atlante Tematico di Topografia Antica* 28: 53-76.

Giorgi, E. 2016. City Archaeology in the Adriatic area: the cases of Burnum in Dalmatia and of Suasa and Ascoli in the Marche regio, in F. Boschi (ed.) *Looking to the Future, Caring for the Past. Preventive Archaeology in Theory and Practice*: 101-123. Bologna: Bononia University Press.

Ferranti, E., Speranza L. 2018. La necropoli altomedievale di Piazza Ventidio Basso ad Ascoli Piceno: prime considerazioni, in E. Cirelli, E. Giorgi, G. Lepore (eds), *Economia e Territorio nell'Adriatico centrale tra tarda Antichità e alto Medioevo*: 405-418. Oxford: BAR.

Pizzimenti, F. 2018. The Urban Archaeology Project in Asculum: the case of Piazza Arringo. *Groma* 3. DOI: 10.12977/groma16.

SOS project: a new challenge for a novel approach to the understanding of an important historical city

S. Campana[1], S. Camporeale[1], J. Tabolli[1], R. Pansini[1], S. Güzel[1],
G. Morelli[3], F. Pericci[4], M. Sordini[4], L. Gentili[5], F. Gianni[5], F. Vitali[6],
G. Carpentiero[7], D. Barbagli[8]

[1] Universitá di Siena (Italy)
[2] Soprintendenza per i Beni Archeologici della Toscana
[3] Geostudi Astier Srl
[4] ArcheoTech&Survey Ltd
[5] LDP Ltd
[6] University of Bologna (Italy)
[7] Soprintendenza Archeologia Belle Arti e Paesaggio del Molise (Italy)
[8] Comune di Siena (Italy)
Stefano Campana - stefano.campana@unisi.it

Nowadays, and from a slightly reductive point of view, it is possible to identify two contrasting types of urban setting. Firstly, there are cities that have been abandoned for so long that they are now defined by such terms as 'former urban areas' or 'one-time townscapes'. Secondly, there are cities that have remained in continuous occupation and which have become major centres of population in the present day ('historical urban landscapes'). The intrinsic characteristics of historical urban contexts are to be identified most obviously in the persistence and *longue durée* of the settlement, in the vertical and horizontal development with the consequent concealment of large parts of their original fabric and in the presence of rich and articulated stratigraphic deposits. These characteristics challenge archaeologists to develop and implement research methods that are clearly different from the study of abandoned cities.

The latter, often corresponding to areas currently used for agricultural activities, are investigated through the analysis of their monumental remains (if present), through archaeological excavation and through surface reconnaissance and aerial photography etc. (Bintliff, Snodgrass 1988). Around the turn of the millennium the discussion within the academic community about possible new approaches was matched by a general improvement within the hard sciences. Among other influences, geophysical prospection began to play a central role. The efficacy of this technique improved dramatically in the 1990s but the authentic revolution has only materialised in the last few years with the application of very large-scale geophysical prospection in once-urban contexts within the Mediterranean area (Vermeulen *et al.* 2012; Johnson, Millett, 2013).

From the very first years of their application, it became clear that the opportunities and potential gains offered by these new techniques and instrumentation were enormous. Large and complex once-urban sites, previously studied for their monumental importance and historical or artistic value through field-walking survey, surface collection and exploratory or targeted excavation, could now be studied in the first instance through non-invasive geophysical prospection, sometimes revealing the entire plan of the town before any intrusive method of investigation was put in hand. This was a truly significant transformation, allowing archaeologists to address specific questions in a way that had not been possible previously. Unsurprisingly, important improvements in the understanding of urbanism followed many of these survey projects. In particular, understanding of urbanism in the Roman Empire benefitted hugely from the integration of remote sensing methodologies in partnership with GIS-based archaeological mapping and of course field-walking survey, artefact-collection and excavation. An important contribution was also made in a variety of cases by aerial photography, both from targeted exploratory flights and through the analysis of 'historical' photographs already available in regional and national archives (Musson *et al.*, 2013). The combined application of these essentially non-destructive techniques has greatly enhanced our knowledge of the scale, structure and chronology of specific buildings and the overall infrastructure within formerly urban contexts, allowing us to look at the wider phenomenon of urbanism from a more valid and comparative viewpoint.

Urban archaeology as we know it today was developed in Great Britain in the period following WWII using stratigraphic excavation as its principal method of investigation, later supplemented by other techniques such as core-sampling and environmental analysis. Today, the most valued techniques in cities still under occupation can be distinguished in several respects from those that are appropriate for deployment in once-urban contexts that have become wholly or partially buried following their abandonment. In ancient cities that have never lost their original urban function one can count on an ample supply of documentary information or archival material which preserves the memory of the past and presents accounts of previous archaeological discoveries. Among the major problems encountered in such contexts, however, there are the inevitable impediments of past and recent urban development. Although in both types of urban contexts, former and historical cities, we have seen very significant progress in our investigative skills, the study of abandoned cities is, on the whole, simpler and has benefited in particular by the possibility of analysing the whole of the archaeological *continuum* (Campana 2018).

Indeed, the systematic presence of buildings, especially those of some antiquity, can through their standing façades, cellars and foundations provide an important source of information, but at the same time they constitute a grave obstacle for the application of many of our standard diagnostic methods, not altogether excluding their use but (for instance) certainly reducing, complicating and slowing down the feasibility of carrying out survey work within the area of the buildings themselves and thereby interrupting the continuity of any overall analyses that can be made. The open areas free of buildings are limited to the often-busy streets, squares, and parks along with a variety of private and public gardens but little else. In these spaces it is equally difficult to deploy many of our usual investigative techniques. The great majority of the public spaces are covered in hard surfaces of one kind or another, precluding direct observation of the underlying deposits and therefore also preventing the use of non-destructive methods such terrestrial or aerial survey, even though this last technique has been shown to have value in certain circumstances and situations. Even the methods of geophysics suffer in the urban context; magnetometry, for instance, struggles in the face of the strong magnetic interference which is inevitable in living towns and cities. Without doubt the most effective method in such contexts falls within the sphere of Ground Penetrating Radar (GPR).

Within this framework we believe that the SoS Project now under way in Siena could serve as a useful pointer to the future. Among Italian historical centres, Siena is one of the least well-known examples from the point of view of the archaeological evidence, notwithstanding several archaeological excavations which are unfortunately almost entirely unpublished to date. However, this paucity of available evidence offers at the same time a significant incentive and challenge to re-address and analyse the 'inhabited urban context' of a vibrant historical city within a truly holistic strategy. The SOS Project (the acronym comes from '*SOtto Siena*', in English 'Beneath Siena'), promoted by the Regione Toscana and coordinated by the University of Siena in collaboration with the Soprintendenza Archeologia, Belle Arti e Paesaggio per le province di Siena, Grosseto and Arezzo (SABAP), in partnership with the Municipality of Siena and Regione Toscana. The project is part of this framework with the aim of overcoming some of the problems and limitations present in the study of cities with long-term continuity of life, responding in particular to the need for a better understanding of the city's ancient fabric and hence to improvements in its conservation. As noted above, for obvious reasons it is not possible in such contexts to analyse the archaeological record with the usually desirable degree of continuity across time and space. However, a clear determination to aim for as complete an integration as possible between the sources and methods currently available within the fields of archaeology and urban studies can allow us to achieve at least a form of knowledge that we could perhaps define as 'intermittent continuity'.

With this in mind the project partners' have identified and developed three main lines of action:

- The development of a 3D Archaeological WEBGIS of Siena shared by the University of Siena, SABAP and the local administration;
- The systematic acquisition of GPR data of all the public spaces in Siena, such as streets, squares, courtyards and gardens that are at least theoretically accessible for the mapping of archaeological features and utilities down to a depth of about 2m and over an estimated 25 hectares of surveyable area (Figure 1);
- GIS data entry of most of the historical-archaeological and geoarchaeological knowledge of the city and of the interpretations of high-resolution radar datasets, allowing the integration and progressive combination with and between the existing documentation and future results from within the city.

The development of a WEBGIS and the data entry of all the known material evidence and information about the city is certainly not an innovative initiative, but the integration with the GPR surveys from all the accessible public and private spaces will make it possible to introduce, to some degree at least, the concept of continuity that is fundamental for the understanding of urban development. This will have substantial implications in the daily activities of SABAP (as well as the Municipality), improving the ability to provide specific means of addressing the needs

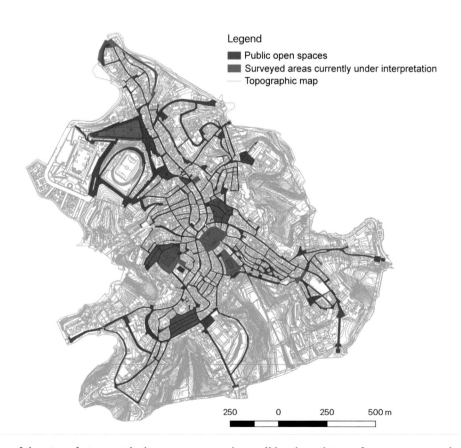

Figure 1. Map of the city of Siena with the open spaces that will be the subject of GPR surveys indicated in red, across a total of 25 hectares in all. Blue shows the areas already surveyed and currently under data processing and interpretation.

Figure 2. Survey activities by IDS STREAM UP radar system.

of sustainable development within the urban area for activities related to entrepreneurship and citizenship in general.

In order to carry out the project, in addition to the existing collaboration between the University and the SABAP, it was necessary to involve other specialized companies and enterprises with differing roles in the research and survey work. From a technical point of view, the management of WEBGIS has been entrusted to the LDP company which, in collaboration with one of the SoS project's Research Fellows, is developing the 3D data viewer. As a fundamental aspect of the project, it should be noted that an agreement has been established with the Superintendency of Rome for sharing use of the framework of the Archaeological Information System of Rome (ArcheoSITAR; Serlorenzi 2019). The availability of a well-structured WEBGIS allows the project to focus its efforts mainly on the development of a 3D visualization system, constituting a powerful tool for interpreting, analysing and managing the inherent complexity of the radar data in a context characterized by a highly complex and closely articulated morphology. The inclusion of the third dimension on such a large scale has never been tested and the impact is not limited to knowledge and research but directly affects the overall evaluation of the risk analysis and the archaeological potential of the city, becoming a substantial tool for urban planning and, more generally, for the management of the subsurface deposits. The acquisition and processing of GPR data is coordinated by Geostudi Astier in collaboration with ATS srl and the project's second Research Fellow (Figure 2). The local administration plays a fundamental role in the provision of technical maps and, notably, in the management of logistical issues of great importance within the urban areas.

The possibility of increasing knowledge on the historical development of the city and presenting the results of the research to the public in the narration of new historical scenarios, both through traditional methods and through immersive technologies, has led to the involvement in the project of the Polo Museale della Toscana, Santa Maria della Scala and the Santa Chiara Lab of the University of Siena, the latter specializing in the development of virtual and augmented reality systems and 3D printing.

The SoS Project, in summary, can be characterized as an ambitious and complex project that from its outset has aimed at methodological innovation in the search for new archaeological understanding of the city of Siena. It has from its first inception involved close collaboration between a wide range of institutions, consultancies and commercial concerns, with the ultimate hope that in time it will be seen as an example that could profitably be replicated in other urban contexts, both within Italy and beyond.

References

Bintliff, J., Snodgrass, A. M. 1988. Mediterranean survey and the city. *Antiquity*, 62: 57–71.

Johnson, P., Millett, M. (eds). 2013. *Archaeological Survey and the City*. Oxford: Oxbow Books.

Campana, S. 2018. *Mapping the Archaeological Continuum. Filling 'Empty' Mediterranean Landscapes.* New York: Springer.

Keay, S., Millett, M., Paroli, L., Strutt, K. (eds). 2005. *Portus. An Archaeological Survey of the Port of Imperial Rome*. Archaeological monograph of the British school at Rome: 15. London: British School at Rome.

Musson, C., Palmer, R., Campana, S. 2013. *Flights Into The Past. Aerial photography, photo interpretation and mapping for archaeology*, Occasional Publication No. 4 of the Aerial Archaeology Research Group in partnership with the ArchaeoLandscapes Europe (ArcLand) Project of the European Union (http://archiv.ub.uni-heidelberg.de/propylaeumdok/2009/, file accessed 24.6.2021).

Serlorenzi, M. 2019. Il SITAR: verso una forma di tutela condivisa in Atti del convegno multidisciplinare di studi 'Forme della tutela' (Roma, 8-9 Giugno 2018), *QdM-Quaderni del Master TPC*, Efesto, Rome 2019.

Vermeulen, F., Burgers, G. J., Keay, S., Corsi, C. (eds). 2012. *Urban Landscape Survey in Italy and the Mediterrenean*. Oxford: Oxbow Books.

Conducting archival research in an interdisciplinary context for Rome Transformed.

F. Carboni[1], E. D'Ignazio[2]

[1] Newcastle University (UK)
[2] Pontificia Università Gregoriana, Rome (Italy)
Francesca Carboni - francesca.carboni6@gmail.com
Emanuela D'Ignazio - emanuela.dig@gmail.com

Archival analysis is one of the tools exploited by the Rome Transformed project to carry out a comprehensive survey of the Eastern Caelian, part of the project's multidisciplinary approach which, alongside scanning and architectural analysis, geophysical surveys and environmental analysis will bring data together to reconstruct the evolution dynamics of this urban sector from the first to the eight century CE.

Data obtained through this type of research contribute to assessing the archaeological and architectural remains still visible in the area and provide evidence of the vestiges that were destroyed or obliterated over the centuries. They play a crucial role in leading structural analysis to a better visualisation of exposed structures and in giving clues for the interpretation of the anomalies detected by the geophysical prospection. Legacy data are also pivotal for defining the reshaping of the landscape throughout the key periods covered by the project.

Our research has been conducted so far through two interconnected actions: the first consisting in the archival, documentary, bibliographical and cartographic survey aiming to obtain valid sources that attest to the existence of archaeological pieces of evidence and to the original features of still standing buildings. The second phase involves cataloguing, recording, and archival data managing, for which we have set up a methodology tailored made for the Rome Transformed project.

Challenges encountered during the first stage of data capture can be easily understood by all the scholars who have dealt with the study of Rome's urban landscape development, in any period, due to the extraordinary rich and intricate archive network of this city.

The distribution of documents in several archives is the direct consequence of the peculiar history of Rome, where for centuries the ruler was the spiritual head of the Church but also the temporal chief of a state (this reflecting in the two largest surviving archives: the Archivio Apostolico – Segreto – Vaticano and the Archivio di Stato di Roma), and where the double role of the city as a spiritual centre and as the capital of the new state was maintained after the unification of Italy in the Nineteen century, with the setting up of the Archivio Centrale dello Stato. The capital of Italy is also the seat of a large and distinct municipal archive (the Archivio Storico Capitolino) and of an equally conspicuous diocesan archive which, unique case among the ecclesiastical archives, has collected almost all the Roman parish archives in addition to the 'Lateranensis' one (the Archivio Storico del Vicariato di Roma).

For the project's purposes, the historical archives of the institutions responsible for the protection of Rome's Cultural Heritage are of particular importance, as well as dedicated libraries, such as the Vatican Library and the Biblioteca di Archeologia e Storia dell'Arte di Palazzo Venezia, both preserving precious manuscripts and original iconographic documents.

Most of the knowledge relating to the archaeological heritage within the Rome Transformed research area is due to the works carried out in the post-unification decades (Figure 1), documented by reports that sometimes lack topographical accuracy. Since the second half of the last century, thanks to increasingly rigorous documentation methods, modern urban planning has been returning abundant data from the subsoil, that are often useful to clarify the information known from the nineteenth-century excavations.

Faced with the huge amount of archive material to be consulted, our research has made significant use of what is now an unreplaceable research tool: the Geographic Archaeological Information System, managed by the Soprintendenza Speciale Archeologia, Belle Arti e Paesaggio di Roma, with the first aim of cultural heritage protection in the urban planning context.

Figure 1. Rome Transformed research area on the Rome's land-use plan (detail) of 1882 (above) and 1908-09 (below). Archivio Storico Capitolino, Cart. 13, 135 e 136.

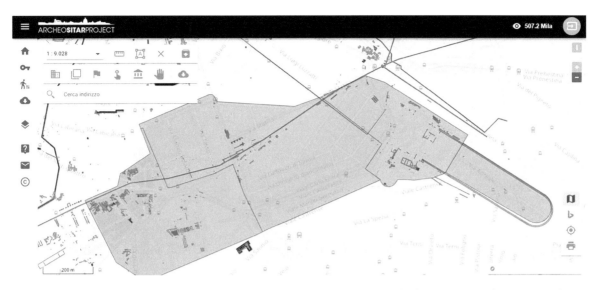

Figure 2. Rome Transformed research area on the ArcheoSitar webGIS platform. Emanuela D'Ignazio from ww.archeositarproject.it/piattaforma/webgis/.

Following an agreement established at the beginning of the Rome Transformed research program, both archive research and the results of the RT project have been interconnected with the ArcheoSITAR project, which is constantly enriched with the updates on new discoveries. So far, we have downloaded from the SITAR platform documentary material relating to about 120 OI (Origins of Information), spread within our area (Figure 2).

Whenever possible, we have increased this amount of data with documents from the current archives of the Sovrintendenza ai Beni Culturali di Roma Capitale and of the Soprintendenza Speciale di Roma, this, in particular, for the area of S. Croce, for which an extensive archival research was already conducted in the past.

Further data acquisition has been undertaken in a variety of ways and concerns many kinds of documents, some of which we acquired as scans or photos, other we could only read, register, by copying text and reproducing images through sketches.

This work, not yet completed, has been conducted according to best practice in topographical research, starting from the bibliographic review of the journals Bullettino della Commissione Archeologica Comunale di Roma (1872-) and Notizie degli Scavi di Antichità (1876-) and going on with the consultation of the precious and abundant documentary material collected in the Historical Archive of the Soprintendenza Archeologica di Roma (ASSAR) in Palazzo Altemps, research facilitated by the archive's recently restyled website (ADA.beniculturali.it).

We had the wonderful opportunity to be involved in the agreement signed between the Archivio Centrale dello Stato and the ArcheoSITAR project, including that invaluable archive source represented by the Gatti Fund, and thus to acquire high-resolution scans of the notes and papers of three generations of excellent archaeologists who recorded the discoveries occurred within the RT area in the span period between the late 19th and the early 20th century (Figure 3).

Collaboration with the Museo Nazionale Romano will, we hope, allow us to digitize the original reports and the extensive number of pictures from the excavations conducted by Santa Maria Valnea Scrinari in the areas of the Ospedale San Giovanni Addolorata, Santa Croce in Gerusalemme and the INPS building in via Amba Aradam, between the 1960s and the 1990s. All of these areas

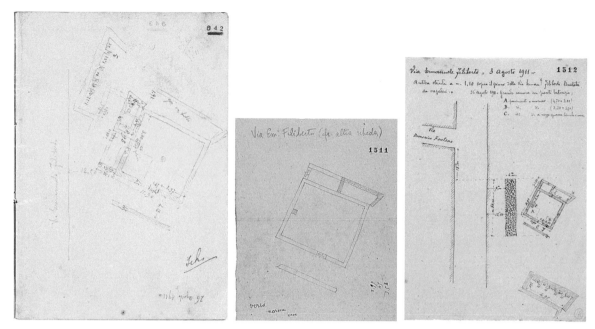

Figure 3. Roman structures discovered in 1911 along Via Emanuele Filiberto depicted, with different accuracy, through three provisional sketches collected in the Archivio Gatti. Archivio Centrale dello Stato, Fondi Gatti Edoardo e Guglielmo, Taccuini, Regio V, 842; Carte, f.6, sf.13, 1511 and 1512).

have been subjected to extensive reappraisal, involving full structural analysis and laser scanning, by the project team.

For the second phase of our research, consisting in data recording, we have tended to conform our methodology to the one conceived by the WebSITAR platform, to profit from such solid experience, based on a long methodological debate, and to facilitate the future interoperability between the two projects. To this end, the multidisciplinary commitment of various members of Rome Transformed led to the definition of an Excel sheet, with fields and vocabulary partly borrowed from the SITAR platform but targeted to the peculiar features of the Rome Transformed project, aimed at the development of a 3D management system of Rome's archaeology and that has resulted fully compatible with the 3D data base tailor-made for Rome Transformed by the LabGEO team. We started as common practice in similar projects to draw up an archaeological map in a CAD environment, positioning, vectorizing and georeferencing all the available information sources useful to describe the transformations of this area between the 1st and the 8th c. CE, based on excavation reports (Figure 4).

An important part of our work, therefore, involved the acquisition and selection of the most suitable cartographic document for the location of the pieces of evidence. Following the principle of regression, usually adopted in the approach to the geographical positioning of historical maps, the cartographic sources were georeferenced from the most recent to the oldest one. The possibility of understanding whether information acquired in distant times and through different investigation systems referred to the same evidence and how much these contributed to its better and more detailed description, relies on the best possible location, in the three dimensions of the information data.

The central role of 3D visualizationto the Rome Transformed project led us to pay particular attention to the third dimension, that is the depth of archaeological structures, sometimes known only through archival documents. Thus, we have tried to assess the ground levels of every detected phase of use for the same structure for different features brought to light during the same investigation (= information source). To get homogeneous data we have been trying to calculate

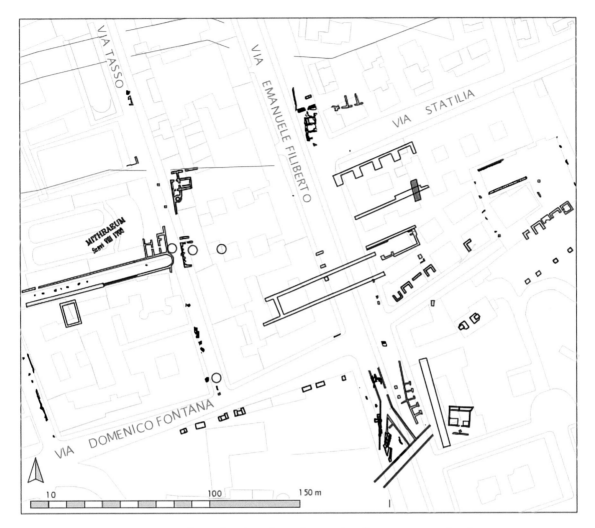

Figure 4. Rome Transformed archaeological map, detail. Francesca Carboni, May 2022.

the absolute value of the heights recorded in the excavation reports as far as it is possible and in the most accurate way, recording the relative heights provided by the original sources (indeed in many cases the indication of the depth of the findings cannot be associated with any reliable benchmark), this approach addresses the twofold requirement of our archive research: source transparency and data comparability.

For ancient excavations and discoveries, we have had to cope with the difference of absolute values reported in pivotal, almost contemporary urban maps, such as the Pianta Topografica di Roma published in 1875 by the Istituto Geografico Militare and Rodolfo Lanciani's *Forma Urbis Romae* (1893 - 1901), in search for possible, alternative altimetry sources.

Beside the well-known difficulties in georeferencing the *Forma Urbis* of Lanciani because of the reasons extensively highlighted by the recent studies on this subject, we check the geometries of the archaeological features there mapped comparing them with the numerical values on handwritten notes, collected in the Lanciani codes, partly still unpublished, at the Vatican Library.

Also, the digitation work on Gatti's sketches consisted in the redrawing of the archaeological features, according to the measures provided on the notes, and in their subsequent georeferencing on the basis of the best contemporaneous and most detailed cartographic support we could find.

The first attempt to compare the structures thus mapped with the preliminary results of the geophysical investigations conducted so far by the various teams within the Rome Transformed project has made it clear how archival data are fundamental for the interpretation of anomalies that have the chance to be correctly located.

Acknowledgements

'Rome Transformed' has received funding from the European Research Council (ERC) under the European Union's Horizon 2020 research and innovation programme (grant agreement no. 835271).

The archive research conducted so far, which has proceeded smoothly despite the restrictions due to the outbreak of the Covid pandemic, has been possible thanks to the availability and support of many people and different institutions.

Access to the documents collected in the historical archive of the Soprintendenza Archeologica di Roma was facilitated in every way by Antonella Ferraro. A special thanks to Alessandra Capodiferro for her patience and to Maria Luigia Attilia for the competent advice in the initial phase of the research.

Collections of documents from the archives of the Soprintendenza Capitolina ai Beni Culturali were provided by Maria Gabriella Cimino, Marianna Franco and Paola Chini, who is responsible for the Archivio Storico e Disegni.

On the side of the Soprintendenza Speciale Archeologia Belle Arti e Paesaggio di Roma, we are particularly grateful to Mirella Serlorenzi, and the ArcheoSITAR team and to Anna De Santis and all the colleagues researching on Santa Croce's area, for the documentation they shared with us and the fruitful discussions.

Finally, sincere thanks to Mirco Modolo, responsible for the Gatti Fund at the Archivio Centrale dello Stato.

References

Attilia, L. 2020. L'archivio storico a Palazzo Altemps: la storia, le carte, il web, i progetti in corso, in A. Pessina and M. Tarantini (eds.), *Archivi dell'Archeologia Italiana. Atti della giornata di studi Archivi dell'archeologia italiana. Progetti, problemi, prospettive* (Firenze, 16 giugno 2016): 33-48.

Baiocchi, V and Lelo, K. 2014. Assessing the accuracy of historical maps of cities: methods and problems, *Città e Storia*, 9, 1: 61-89.

Bosman, L., Haynes, I. P. and Liverani, P. (eds) 2020. *The Basilica of Saint John Lateran to 1600*. Cambridge: Cambridge University Press.

Colli, D., Martines, M.T. and Palladino, S. 2009, Roma. Viale Manzoni, Via Emanuele Filiberto. L'ammodernamento della linea A della Metropolitana: nuovi spunti per la conoscenza della topografia antica, *The Journal of Fasti Online* 154.

Colini, A.M. 1944. *Storia e topografia del Celio nell'antichità*. Rome: Tipografia Poliglotta Vaticana.

Colli, D. 2020. Costantino, il Sol Invictus e il palazzo Sessoriano di Roma. Spunti, dati e considerazioni per una ricostruzione della residenza imperiale. *Journal of Ancient Topography - Rivista di Topografia Antica* 30: 255-296.

Consalvi, F. 2009. *Il Celio orientale. Contributi alla Carta archeologica di Roma, Tavola VI, settore H*. Rome: Quasar.

Coarelli, F., Modolo, M., Serlorenzi, M., Jovine, I., Lamonaca, F. 2018. Il progetto di valorizzazione dell'Archivio Gatti, in E. Lo Sardo and M. Modolo (eds), *Nuove fonti per la storia d'Italia. Per un bilancio del secolo breve*: 186-192. Rome: De Luca editori.

Lelo, K., Morelli, R., Sonnino, E., Travaglini, C. M. 2002. GIS e storia urbana, in R. Morelli, E. Sonnino, C. M. Travaglini (eds), *I territori di Roma: storie, popolazioni, geografie*: 191-211. Rome: Università degli Studi di Roma La Sapienza.

Serlorenzi, M., Leoni, G., Jovine, I., De Tommasi, A., Varavallo, A. 2016. Il Sistema Informativo Territoriale Archeologico di Roma. Processi, metodi, strumenti e contenuti per l'informazione archeologica sul web, in A. Scianna, M. L. Scaduto (eds), *Atti del XV Meeting degli utenti italiani del GRASS e GFOSS* (Palermo, 12-14 Febbraio 2014), *Bollettino della Società Italiana di Fotogrammetria e Topografia*, 4: 56-62.

Severino, C. G. 2019. *Roma. Esquilino 1870-1911*, Rome: Gangemi.

The challenge for archaeologists using geophysics in urban areas

M. Dabas[1], F. Blary[2], G. Catanzariti[3]

[1] CNRS, UMR8546, Ecole Nationale Supérieure, Paris (France)
[2] CReA Patrimoine, ULB, Bruxelles (Belgium)
[3] 3DGeoimaging, Turin (Italy)
Michel Dabas - michel.dabas@ens.psl.eu

Introduction

Geophysical prospecting, even if now used routinely for managing cultural heritage questions in open areas, suffers in currently urbanized environment from various difficulties which makes its use non-trivial. A certain number of constraints in the urban environment are generally well known: the presence of mechanical vibrations, the electromagnetic noise, the presence of infrastructure, of furniture, of pedestrians, of cars above the surface, the presence of underground modern services (water, gas, electricity, telecoms, etc.), and most obviously, the presence of above ground structures which tend to fragment into smaller spaces the areas that can be measured. Others like the very high heterogeneity of the subsoil (rubble and a highly complex multi-phase stratigraphy) remains the biggest challenge for the geophysicists and ultimately for archaeologists during the interpretation process.

The very good results obtained over abandoned Roman towns for example are due to the conjunction of facts that are seldom encountered in today's city centres: low depth of structures (< 1 m), existence of a destruction phase that has cleared up the rubble and made apparent walls or their negative counterpart, and/or the existence of a single phase of construction of the urban layout. The situation of today's cities is closer to the one of 'tells' in middle-east countries.

If the first archaeo-geophysical surveys in cities were very focussed and were often limited to the study of religious buildings or parks, in recent years the appearance of new methods such as the electrostatic method (also called CCR – Capacitively Coupled Resistivity) or motorized ground radar (GPR) allow the investigation of very large areas. Two examples are developed: the research conducted in Alexandria twenty-five years ago and a newer one not yet published, carried out in 2018-2019 in Brussels.

Case study #1: Alexandria, Egypt

This case study represents the first large experiment done in a modern city. Following an idea of J.Y. Empereur in 1995, we (Albert Hesse *et al.* 2002) were looking for traces of the old causeway (Heptastadium, a causeway of 1155 m long) described by Strabo and others (Aristo, Plutarch, Seneca) between the old town of Alexandria and the island of Pharos in Egypt. This causeway is sometimes described as an embankment (χῶμα) that could form a bridge (with piles?) and an aqueduct which connects the old town to the island of Pharos. Before our work there was only one hypothetical route that was proposed by Mahmoud Bey Al-Falaki in the last century and situated halfway between the two banks of today isthmus. But looking at topographic features (orientation and width of the old roads, parcels in the 1930 cadastre, altitudes and the distribution of sewers), Albert Hesse proposed another route closer to the west bank of the isthmus. This hypothesis focused all the geophysical work in this region and four methods were used in 1997 (Seismic refraction, EMI, GPR and Electrostatic imaging) mainly along six modern roads (approximately 100m apart) crossing the hypothetical trace of the Heptastadium.

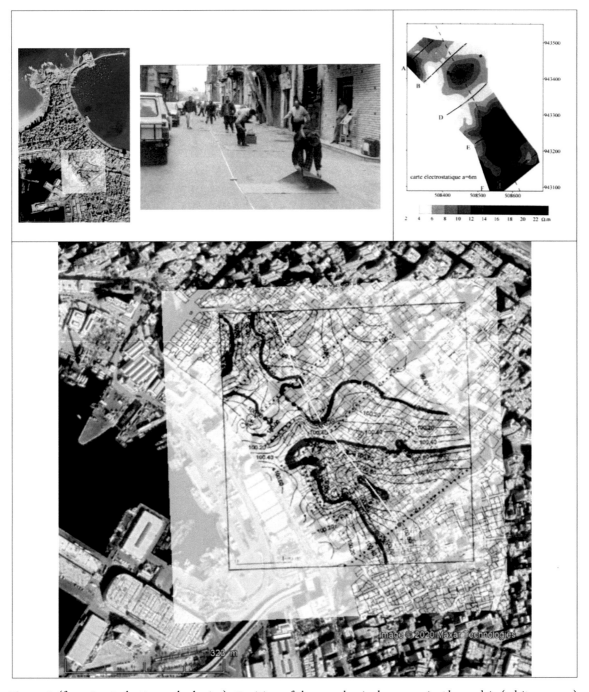

Figure 1. (from top to bottom, clockwise): Position of the geophysical surveys in Alexandria (white square); Electrical Profiling using independent electrostatic quadripoles in Wenner configuration (a=6 m) along one of the streets (credit: A. Hesse); Map of apparent electrical resistivities obtained by krigging the data along the 6 profiles (Hesse *et al.*, 2002); Superposition of microtopography and Google Earth showing in yellow the proposed causeway for the Heptastadium (credit: A. Hesse, M. Dabas).

We will not discuss in detail (cf. Hesse *et al.*, 2002) the results of these geophysical profiles in a very difficult environment (very low resistivities due to the presence of salted water, decompacted sediments, high industrial noise). Seismic refraction and GPR were not successful but Electrostatic data and to a lesser extent conductivity data (EMI) made possible the delimitation of resistive bodies at a depth of roughly 4 meters.

Nevertheless, it is clear, as was expected since the beginning of this work, that the Heptastadium does not constitute a perfect linear geophysical anomaly (Figure 1). However, the two methods show that there exists, in the studied sector, a slight eminence aligned in the right direction and composed of resistant materials contrasting with more loose alluvium on the sides, validating the hypothesis of some 'openings' across the Heptastadium for communications between the western and the eastern harbour. These geomorphological issues together with micro-topographical issues were important for strengthening the hypothesis of a causeway in the western part of the isthmus.

Case study #2: Brussels, Belgium

The program '*Voir sous les pavés*' (Seeing under cobblestones) launched in 2018 for the study of the Grand-Place in Brussels, under the direction of François Blary (ULB) and Michel Dabas (CNRS-ENS), aimed to detect and map the archaeological structures under this square and nearby streets without performing excavations. It corresponds in the case of Brussels to a set of large-scale integrated non-destructive surveys that prefigure what could be undertaken in other downtown areas. This project is part of a larger project launched in 2017: the BAS (Brussels Archaeological Project). It aims at finding the traces of the Brussels of the Middle Ages of which nothing remains on the surface. This research focuses on the study of all accessible underground structures and in particular cellars within the oldest area of Brussels known as the 'Pentagon'. Grand-Place occupies the central position of this Pentagon.

Figure 2. MP3 prototype for measuring apparent resistivity, operators (A. Tabbagh and S. Flageul) are within a 'moving' safety zone delimited by yellow tape; Robotic total Station operated by B. van Nieuwenhoeve; GPR (StreamC 600 MHz and StreamX 200 MHz, IDS GeoRadar Corporation) G. Morelli and G. Catanzariti; Vertical electrical Sounding between the cobblestones in the Townhall (A. Tabbagh).

Methodology

The current appearance of the Grand-Place is close to that of 1695, after the bombardment by the French. Before 1695, little is known: the square was probably much smaller, texts mention buildings but their location are unknown. One or more fountains existed and are depicted in some drawings. Also, more ephemeral structures have existed: Gas streetlamps, places for Christmas trees, market place, and podium for public executions. All these remains could eventually be detected by geophysics. Only a single very small archaeological excavation has occurred near the Town Hall and some occasional discoveries like a water well or a cave are known.

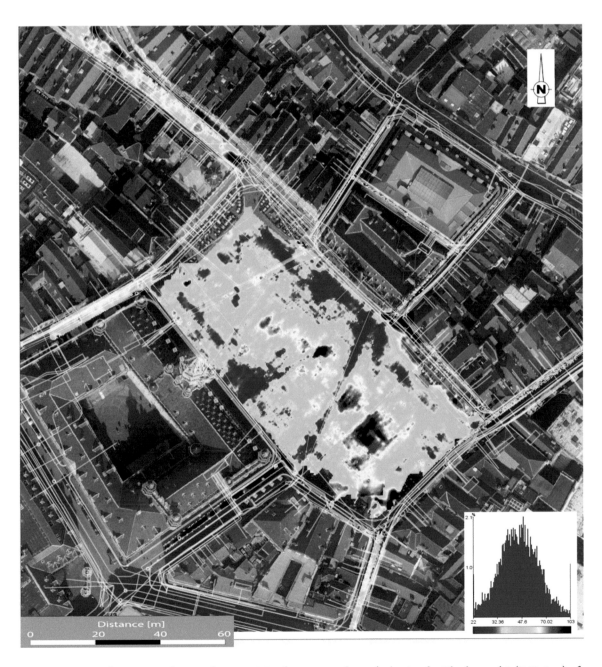

Figure 3. Map of apparent electrical resistivities (22 to 103 Ohm.m) obtained with channel 2 (DOI: 1 m) of the MP3 prototype (charge coupled resistivity meter) overlaid on 5 cm resolution orthoimagery (June 2016, source: cirb.brussels, Open licence) and utilities (in white). Processing and source: M. Dabas, www.chronocarto.eu [06/2021].

Figure 4. 200 MHz GPR amplitude time-slice centred on 0.22m overlaid on ortho-imagery (June 2016, source: cirb.brussels, Open licence) and utilities (in blue). Processing and source: G. Catanzariti, www.chronocarto.eu [06/2021].

Before the surveys took place, it was possible to get a high definition (5 cm) aerial ortho-image and a map of the different utilities. But no georeferencing of these utilities existed and one of the purposes of the survey was also to map all utilities under the main square and the streets around.

Four classes of methods were used simultaneously in the Grand-Place: ground penetrating radar (GPR), electrostatic method (also named CCR - Capacitively Coupled Resistivity-), Vertical Electrical Sounding (VES), and micro-topography (Figure 2). The biggest challenge was to get authorizations to survey this UNESCO-square full of tourists, with security issues (2016 suicide bombings in Brussels), with heavy car traffic during the early morning when shops are stocked, and with a high tavern activity in the afternoon and at night.

Several 3D-multi-channel radars (Stream X and C) were set up (Gianfranco Morelli, Geostudi Astier) to find a compromise between depth of investigation, spatial resolution and measurement speed. Simultaneously, a new prototype of electrostatic resistivity-meter 'MP3' specifically developed for this project was implemented (S. Flageul and M. Dabas). It allows the measurement of electrical resistivity simultaneously for three depths of investigation down to 2 meters. Positioning was done by a robotic total station which proves to be very effective in such a situation (compared to GNSS) even if it has to be moved several times during the survey of surrounding streets and has

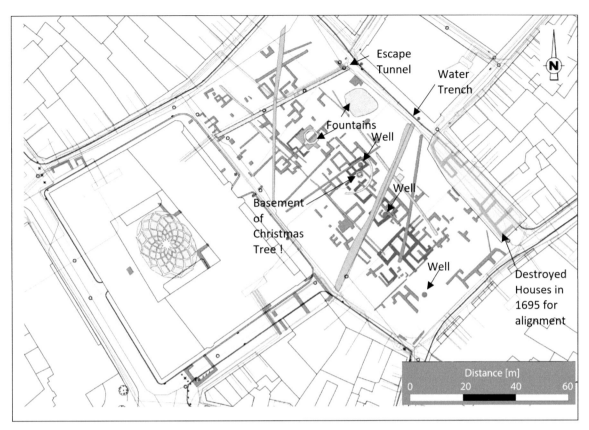

Figure 5. First joint interpretation of resistivity maps and GPR time-lapse maps. Processing and source: F. Blary [03/2019].

to be guarded all day by an operator. It was used for all geophysical methods and consequently all data set were spatially coherent, saving lot of efforts during post-processing phases. The post-processing of these data made it possible to calculate a digital model of the soil surface (DSM) with a post-registration into the Belgian cartographic system (Lambert72) using some GPS control points.

The relatively simple processing and interpretation of the CCR data in a GIS has to be opposed to the very complex processing and interpretation of 3D-radar data: radar maps in the form of time-slices, using a representation of the amplitude (phase) or the magnitude (envelope), taking into account - or not - the topography was tested.

Results

Resistivity (Figure 3) was very successful and able to detect several resistive structures that can be interpreted as the rest of old dwellings, and despite the signature of some utilities, generally depicted as conductive elongated anomalies, distinctive walls of houses were detected all over the Grand-Place. These anomalies were detected with the most superficial quadrupole (a=0.5 m), meaning that these anomalies are superficial. They are best seen with the 1m dipole and have been confirmed with the different maps corresponding to the two orientations of the dipoles. The highest gradient corresponding to probably a massive building can be seen in the south-east corner (Figure 5). The density of resistive anomalies is less in the western part of the Grand-Place. We think that this does not necessarily imply a lower density of constructions in this area, but perhaps deeper or more erased structures.

The phase of destruction and reconstruction of the XVII c. Grand-Place was clearly picked up especially in the South-Eastern part: the widening of the place to the east has resulted in a re-alignment of the houses that resulted in the destruction of the former front of the houses (pushed approximately 10m to the North-East).

Among the known utilities, only one was clearly picked-up within the CCR data and corresponds to the trench associated to a water pipe. The other utilities were too small to be imaged by resistivity but were picked up by GPR.

GPR results, despite many maps corresponding to the different frequencies used, to different processes, and even to different polarizations, were hard to interpret. Unfortunately, Radar penetration even at 200 MHz was low (1m approximately). We have often encountered this situation and it is a drawback when using GPR in downtown areas. For Brussels, this can be explained by the low electrical resistivity measured both by CCR and confirmed by a single Vertical Electrical Sounding between the cobblestones.

Nevertheless, the GPR results demonstrated, beside a very high number of unknown utilities, the existence of the basement of two fountains and probably three water wells (Figure 4) which were not picked-up in the CCR data. The spatial resolution of GPR was necessary to detect these small-scale anomalies (compared to resistivity imaging). As an example of known structures, the place of the basement of the Christmas tree and the one of former water wells were discovered in the time-slice GPR maps (but were not known before our survey).

Surprisingly, the signature of the dwellings and walls discovered with resistivity imaging was not so clear in GPR maps. Only GPR amplitude time-slices (not envelope data) was able to delineate these limits thanks to a difference in texture and we should admit that the confrontation with ER maps was mandatory to secure our interpretation.

We should acknowledge that, without the information gained with old drawings and paintings, we will probably have missed some of the archaeological structures. We believe that the good results of this project is due to the integration of the archaeologists at every stage of the process. And of course, the archaeological interpretation of geophysical data from Grand-Place is a challenge which is not yet finished. All results will be published in a book, in a public leaflet and through a web-GIS.

Conclusions

Through these two examples, we have shown how we have tried to answer the three main questions dealing with geophysics in urban spaces: how is it possible to undertake multidisciplinary research in large, urbanized settings; how can we have managed the acquisition, processing and integration of large amount of data; and finally, how can we have interpreted and disseminated the results of such urban-scale projects?

In both examples, it was mandatory to use several classes of geophysical methods simultaneously to solve the archaeological questions: the presence of the Heptastadium in Alexandria, and the presence of medieval buildings in Brussels. But also, the ancillary role of micro-topography was highlighted for both studies. The role of the electrostatic method (CCR) as a quick mapping technique was demonstrated even if the geometry of such a system prevents it being used easily to investigate deep deposits (>3m). Processing CCR data is also very easy compared to processing GPR data. The high horizontal and vertical resolution of GPR data is nevertheless important to resolve small scale anomalies, define sharp limits of structures and ultimately interpret structures at a stratigraphic level.

Processing of data and integration of all types of data is also very important and is still a challenging part of this work. Our interpretation heavily relies on these processing workflows that can – and should- evolve during the interpretation process. Our experience shows that without extra information like old plans and drawings, our geophysical interpretation may remain very poor.

Finally, we can ask ourselves what exactly is the role of geophysical information obtained in such contexts? Is it only useful for archaeologists, or can we imagine a more integrated approach with geotechnical engineers, network managers, urban planners? Can it be used blindly as in the case of an archaeological diagnosis or does it have to answer specific questions? Even if it is still an open question, our experience and certainly the Rome Transformed project are building new bridges in our knowledge.

References

Hesse, A., Andrieux, P., Atya, M., Benech, C., Camerlynck, C., Dabas, M., Fechant, C., Jolivet, A., Kuntz, C., Mechler, P., Panissod, C., Pastor, L., Tabbagh, A., Tabbagh , J. 2002. L'Heptastade d'Alexandrie, in J.-Y. Empereur, *Alexandrina 2*. Etudes Alexandrines 6: 191-273. Cairo: IFAO.

Department of Parks and Wildlife, 2001, Department of Parks and Wildlife, Canberra, Shipwreck inspection, viewed 18 October 2022 https://www.chronocarto.eu. Dabas and Blary, Le projet Grand-Place de Bruxelles, accessed [10/18/2022] private access, should become public in 2023 or 2024.

Methods and techniques for the interpretation and reconstruction of the ancient landscape outside the Aurelian Walls

E. Demetrescu[1], C. Gonzalez Esteban[2], S. Morretta[3], R. Rea[3]

[1] CNR-ITABC, Istituto per le Tecnologie Applicate ai Beni Culturali, Rome (Italy)
[2] University of Southampton (UK)
[3] Soprintendenza Speciale Archeologia, Belle Arti e Paesaggio di Roma (Italy)
Emanuel Demetrescu - emanuel.demetrescu@cnr.it

Introduction

This paper presents methods and techniques for the interpretation and reconstruction of the ancient landscape outside the Aurelian Walls, between the Circus Varianus and Porta Metronia. The majority of data used is derived from the preliminary investigations for the construction of the new Metro Line C in Rome. This area of the city changed profoundly at the beginning of the 20th-century as a result of a series of landfills for the construction of the new Appio-Latino district. These changes led to the disappearance of an important waterway, the Marrana, Rome's third-largest river after the Tiber and the Aniene. Targeted investigations utilizing coring, archaeological excavations and the examination of archival documentation (Demetrescu *et al.* 2011) over 14 years have led to a profound understanding (Rea 2011, Morretta and Rea 2018, Morretta and Rea 2020) of a sector of the city that was poorly known.

Digital applications have been a tool for the integration, study and management of archaeological and historical landscape data (Demetrescu 2020). The methodological choice that most influenced the research was to provide a complete representation of the data in a 3D environment (within computer graphics software) allowing optimal visual coherence and the possibility of comparing and integrating groups of different sources.

Interpreting the Ancient landscape: methods and techniques

It is not possible here to describe the complete results that have been obtained thanks to the integrated study of the aforementioned topics using 3D digital tools. Only three main aspects will be taken into account to exemplify the potential for the reconstruction of the ancient landscape: due to the extensive reconstruction of the large urban stratigraphies and in particular of the 'non-anthropic' interface (i.e. areas untouched by human interference), it has been possible to understand the dynamics of the construction of the Aurelian Walls, built following a massive burying activity of part of the original bed of the Marrana and the displacement of the watercourse of the same towards the south, in the area that goes from the current gardens of via Sannio to Porta Metronia (Figure 1). The second topic concerns the depth of the Roman inhabitation below the walking level of the present town and the assessment of the impact of the underground tunnels within the archaeological potential. The third topic concerns the validation of a methodology for the reconstruction of large urban stratigraphies from geognostic data and in particular from the archaeological interpretation from corings (Demetrescu and Fontana 2011).

During the planning of the underground line C, it became necessary to understand the stratigraphy in this section of the city with the precise aim of protecting the buried archaeological heritage. Given the enormous extension of the area under investigation, a 'sampling' approach was carried out through core drilling campaigns, which were subsequently subjected to archaeological interpretation, using a standardised scheduling apparatus for the project (Demetrescu and Fontana 2009). Each core drill had two main reference points, the top at the walking level of the city and, at

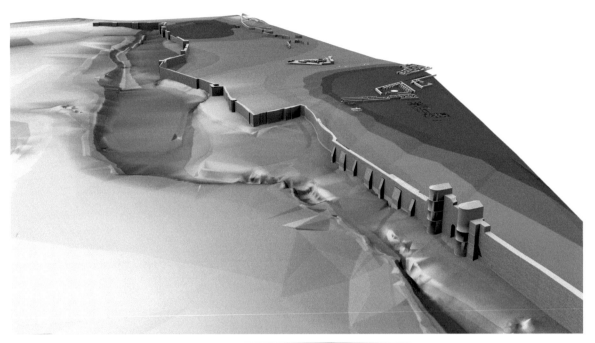

Figure 1. Altimetrical reconstruction plan of the non-anthropic landscape with ideal superimposition of the future Aurelian Walls; below, the landscape of the late Roman period following the construction of the walls themselves. (Image: Emanuel Demetrescu)

its lower limit, the beginning of the so-called 'non-anthropic' terrain. This interface coincides with the 'sterile' terrain, in our case almost exclusively composed of blue/yellow clays or the tuffaceous geological stratigraphy. Between these two extremes, floor levels, the tops of wall structures or the lowest points of building foundations were found. In other cases, large subterranean stratigraphies composed of ceramic material and compacted building remains were found. In addition to the mapping of the elevation points of the various interfaces (non-anthropic, Roman, late antique), the

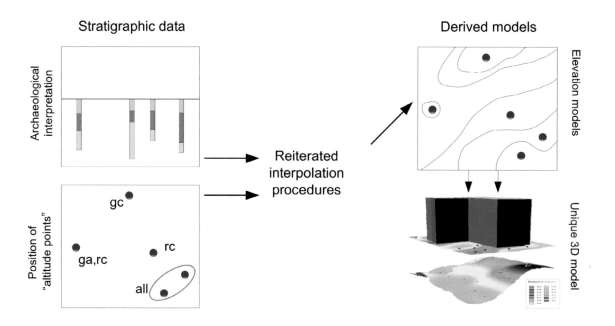

Figure 2. Schematic example of the creation of a stratigraphic interface through the interpolation of height points. (Image: Emanuel Demetrescu)

planimetric documentation with contour lines at 1m of 1906 was added. The re-elaboration of this information has given rise to an original methodology for the creation of height contour models related to the different phases of the life of a landscape from heterogeneous data (Demetrescu *et al.* 2011, Figure 2). In particular, these three 'families' of documents were involved:

1. historical cartography;
2. excavations and known finds from archaeological literature and archival documents;
3. archaeological readings from continuous coring.

The production of digital models of the large urban stratigraphies resulted in an 'expert system' in which the 3D representation of the known archaeological excavations and the metro project (stations, underground galleries, shield introduction shafts) were also integrated. This dataset was used to obtain derived two-dimensional sections useful for the evaluation of the project's impact on archaeology. These drawings facilitated the modifications that were necessary for the design and provided a visual tool to satisfy the authorisation process by the competent authorities. The database was implemented within a GIS software (Grass) so as to bring together the tabular data relating to the cores and the cartographies and artefacts of the engineering works from the CAD design environment. In addition to this GIS database, a three-dimensional 'mirror' representation was maintained within computer graphics software (Blender 2.49). Sections were also elaborated to estimate the number of cubic metres to be excavated for each expected period (modern, medieval and Roman basements; Roman dwelling levels; etc.) with the consequent possibility of quantifying time efforts and methods of the worksite, *e.g.* the number of workers, type of skills and general workflow (Fontana *et al.* 2011).

Once an elevation model of the so-called non-anthropic phase was obtained, it was possible to represent this part of the city at a time before the urbanisation phenomena (an 'ideal' phase at 1000 BC). In this case, the natural environment was the main content taken into account. The territory was divided into ecosystems (riverbed, areas near the water, plateaus, slopes and escarpments) and associated with the characteristic tree species for the Roman area in that specific period (made in Blender 2.5).

During the excavation of the station buildings and of the shield introduction shafts, the quantity and complexity of the available data increased thanks to the stratigraphic excavations. The landscape, already investigated in previous years, was thus enriched with precise information. These discoveries offered the precious possibility of validating the reconstructive models already obtained through coring and, consequently, the methodology that produced them (the stratigraphy found in the excavations has confirmed what had been hypothesised on an urban scale). The quantity and quality of the stratigraphic record allow us to develop reconstructive hypotheses with a greater degree of reliability. In particular, on the building structures brought to light in the archaeological excavations, it is possible to highlight the structural gaps due to the changes that have occurred over time. In the next section, an example of stratigraphic reconstruction through the Extended Matrix will be presented.

Case study: stratigraphic reconstruction of the Amba Aradam Rooms 14 and 15

Amba Aradam Rooms 14 and 15

The site is located in Rome, Italy, on the southern slope of the Caelian, forming part of a neighbourhood reserved for the *castra* during the first half of the 2nd-century AD (Morretta and Rea 2020).

The Amba Aradam station archaeological project was active between 2015 and 2018 covering an area of 3000 square meters. The structures, found at depths of 9 and 12 meters, have been interpreted as a military complex, consisting of a long building with 39 rooms facing an intermediate corridor (soldiers' quarters) and two perpendicular buildings (the so-called 'House of the Commander' and a service building). These elements formed part of a military complex (*castra*) dating back to the Trajanic Era and completed by emperor Hadrian during the first half of the 2nd-century. The building would still have been in use until the second half of the 3rd-century A.D., probably a seat of the urban militia. It was finally abandoned and stripped of all reusable materials in the 3rd-century AD during the construction of the Aurelian Walls.

Rooms 14 and 15 are found on the westernmost side of the excavation and have been regarded as common service areas (Morretta and Rea 2020), in use at least since the Trajanic Era and up until the 3rd-century AD

This area was chosen to test the method proposed to reconstruct, source and visualise the great complexity of urban archaeology in a simple way.

Methodology of the Amba Aradam Case Study

The reconstruction of the Amba Aradam Rooms 14 and 15 followed the workflow proposed to create a source-based virtual model with the EM language (Figure 3), which can be divided into:

- Acquisition of data,
- Creating the hypothesis,
- Creating metadata and paradata,
- Modelling the textured reconstruction,
- Creating and storing results/outcomes.

The fundamental records needed for this workflow were the Harris Matrix and internal reports from the excavation and the reality-based model (photogrammetry) of the remains found, one per period of occupancy if possible.

With this information, before modelling commences, a hypothetical reconstruction is proposed using finds, comparative examples, books and general rules of Roman construction. Furthermore,

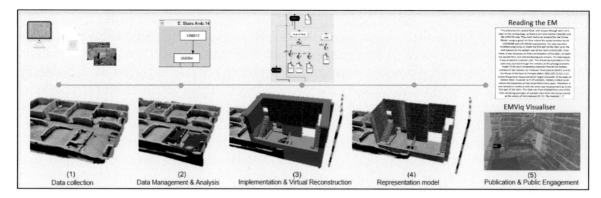

Figure 3. Graphic composite of the specific methodology and results achieved in the Case Study of the Amba Aradam Rooms 14 and 15: a. photogrammetry model; b. Study of the remains; c. Creation of the model and its metadata and paradata; d. textured reconstruction model; e. results and outcomes: visualisation of the model within EMViq and reading of the reconstruction process through the EM. (Composite: Cristina Gonzalez-Esteban)

all the sources (metadata) and the chain of thoughts/actions (paradata) developed to create the hypothesis were recorded in a '*Dossier Comparatif*' format (Demetrescu 2021). This information is joined through the modelling of a Proxy, done using Blender 2.70, which contains the link to the EM and therefore, the metadata and paradata of the reconstruction process (Figure 4).

Once the proxy was approved a texture source-based model based of the aforementioned was made using Blender 2.90.

The linking of the metadata and paradata was developed at the same time as the modelling allowing a review of the reconstruction during the process. Using yEd software (https://www.yworks.com/products/yed), the Harris Matrix was 'completed' with the information of the 'Dossier Comparatif' reshaped in different kinds of nodes based on the reliability of their source, creating what is called an Extended Matrix (Demetrescu 2015). This information was then linked to the model through the usage of the EMtools Blender plug-in, completing the proxy model (Demetrescu and Ferdani 2021:19). Each proxy asset is joined with the US or USV node from the created graph, and therefore, completing the proxy model with its metadata and paradata.

Results and Outcomes of the Amba Aradam Case Study

The final result of this project allowed the creation of a source-based model of Rooms 14 and 15 from the Amba Aradam Roman Barracks. The outcomes produced included a textured model of the site and a sourced proxy model, which were uploaded into the still-developing EMViq visualiser (Fanini and Demetrescu 2019). The latter allows interaction and querying of the uploaded source-based model through a desktop interface or VR, showing the temporary development and history of the rooms. This aims to make the research data and the work done on this site accessible for future research and public engagement of the history of the Roman Barracks found at the Amba Aradam Station from the Metro Line C in Rome.

Beyond the visually appealing results achieved, the most important aspect that validates the creation of a 4D source-based reconstruction is its capacity to ease the reading of the timespan story of the site and the new history that the archaeological works are creating to make the proposed hypothesis.

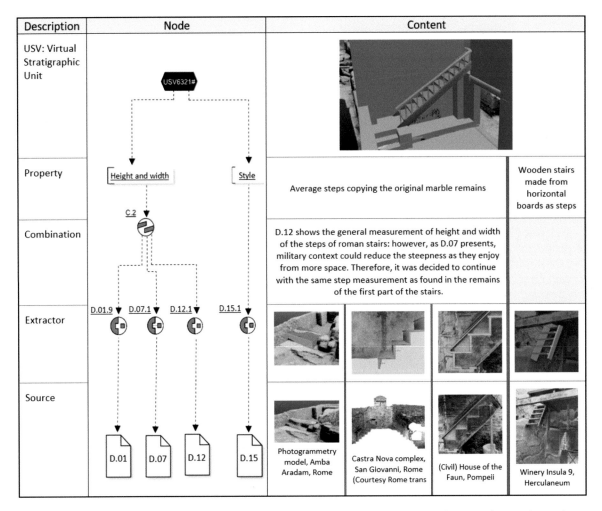

Figure 4. Graphic exemplification of the USV validation process, showing how the metadata and paradata of the reconstruction is built through the Extended Matrix. The example presented of the wooden stairs on the Eastern and Western walls of Room 14 leading to the second floor. (Graph: Cristina Gonzalez-Esteban)

Conclusions

This contribution focused on the definition of methods and techniques to integrate different approaches at different urban scales involving different sources connected by a *fill rouge*: the stratigraphic record. From the landscape to the single building, the EM formal language has proven to be a straightforward method to document the complex process of reconstructing the evolution of an archaeological site, specifically, the lifespan of rooms 14 and 15 from the Roman barracks at the Amba Aradam Metro C station. The project is still under development, especially the modifications in the EM language in order to better fit the requirements of the landscape's record.

Acknowledgements

The presented case study of the Amba Aradam Rooms 14 and 15 was carried out within an assignment at the CNR-ISPC and was partially supported by the award by the La Caixa Foundation of a scholarship towards the Masters studies of one of the authors.

References

Demetrescu, E. 2015. Archaeological stratigraphy as a formal language for virtual reconstruction. Theory and practice. *Journal of Archaeological Science* 57: 42-55.

Demetrescu, E. 2020. La ricostruzione dei paesaggi antichi attraverso l'integrazione di dati, metodologie e tecniche: presente e futuro dagli esempi della Metro C a Roma, in *Archeologi nelle terre di bonifica. Paesaggi stratificati e antichi sistemi da riscoprire e valorizzare*: 135-151. Padova, Italy: TerrEvolute, 2.

Demetrescu, E. 2021. Extended Matrix Handbook: La trasformazione del record archeologico in ipotesi ricostruttiva. Internal Project Report. Unpublished.

Demetrescu, E., Ferdani, D. 2021. From Field Archaeology to Virtual Reconstruction: A Five Steps Method Using the Extended Matrix. *Applied Sciences* 11 (11). DOI: 10.3390/app11115206.

Demetrescu, E., Fontana, S. 2009. Archeo-restituzioni territoriali e urbane, valutazione del rischio archeologico e software open source. *Archeologia e Calcolatori* 2: 95-106.

Demetrescu, E., Fontana, S. 2011. Metodologia di restituzione delle stratigrafie archeologiche sepolte, in R. Rea (ed.) *Cantieristica archeologica e opere pubbliche. La Linea C della Metropolitana di Roma:* 111-120. Milan: Electa.

Demetrescu, E., Fontana, S., Rea, R. 2011. Conoscenze pregresse e nuovi dati: l'evoluzione del paesaggio, in R. Rea (ed.) *Cantieristica archeologica e opere pubbliche. La Linea C della Metropolitana di Roma:* 61-89. Milan: Electa.

Fanini, B., Demetrescu, E. 2019. Carving Time and Space: A Mutual Stimulation of IT and Archaeology to Craft Multidimensional VR Data-Inspection, in A. Luigini (ed.) *Proceedings of the 1st International and Interdisciplinary Conference on Digital Environments for Education, Arts and Heritage. EARTH 2018. Advances in Intelligent Systems and Computing, vol 919*: 553-565. Cham: Springer Nature Switzerland. DOI: https://doi.org/10.1007/978-3-030-12240-9_58

Fontana, S., Marton, S. Rea, R. 2011. Stazione San Giovanni, in R. Rea (ed.) *Cantieristica Archeologica e Opere Pubbliche: La Linea C della Metropolitana di Roma:* 135-176. Milan: Electa.

Morretta, S., Rea, R. 2018. Una nuova caserma alle pendici meridionali del Celio, in A. D'Alessio, C. Panella, R. Rea (eds) *I Severi. Roma universalis L'impero e la dinastia venuta dall'Africa:* 190-199 Milan: Electa.

Morretta, S., Rea, R. 2020. *Roma. Una caserma alle pendici del Celio (II sec.): gli alloggi dei soldati, la domus del comandante, il giardino e l'edificio di servizio,* in C. Wolff, P. Faure (eds) *Corps du chef et gardes du corps dans l'armée romaine:* 387-410. Lyon: CEROR.

Rea, R. (ed) 2011. *Cantieristica archeologica e opere pubbliche. La linea C della metropolitana di Roma.* Milan: Electa.

The archaeological area of S. Croce in Gerusalemme: new data for the reconstruction of the ancient landscape

A. De Santis[1], L. Bottiglieri[2], D. Colli[1], C. Rosa[3], M. Solvi[1]

[1] Soprintendenza Speciale Archeologia, Belle Arti e Paesaggio di Roma (Italy)
[2] 'Sapienza' Università di Roma (Italy)
[3] SIGEA Lazio; Istituto Italiano di Paleontologia Umana (Italy)
Anna De Santis - anna.desantis-01@cultura.gov.it
Laura Bottiglieri - laura.bottiglieri@uniroma1.it
Colli Donato - colli.donato@tiscali.it
Carlo Rosa - carlorosa62@gmail.com
Marco Solvi - marcosolvi@gmail.com

The archaeological area of S. Croce in Gerusalemme, an Imperial residence during the Severan age and the Constantinian period, has recently undergone several restoration projects and archaeological investigations. The evidence that emerged from these interventions, alongside a thorough data review of past research, has allowed the re-assessment of the buildings that originally belonged to the Imperial complex and a new understanding of the construction phases, as well as the residence's original extent and internal articulation. The study of historical cartography and the analysis of stratigraphy based on the drilling of cores reveals that in many cases planning decisions were conditioned by the geomorphology of the site, which – marked by deep ditches and steep differences in height – constrained the positioning of the buildings in the area. We are trying to reconstruct the historic evolution of this area and, also through three-dimensional graphic models, to suggest a new and in-depth reading of the complex, both from a structural and chronological point of view.

Survey methodology

The 2015 excavation , conducted inside the apse of the Civil Basilica (so-called Temple of Venus and Cupid), documented through a photogrammetric survey, launched a campaign of investigations and surveys that has extended to the buildings of the Constantinian domus, to the portion of walls and aqueduct to the north east of the area. The graphic documentation of excavation, performed by means of photogrammetric survey of the removed stratigraphy, allows us to connect the data of the investigations carried out in different campaigns and to obtain three-dimensional levels of the archaeological layers which can be related to other findings to recompose in a 3D environment a model of the volumes of stratigraphy integrated with the structures and finds.

Our survey methodology mitigates, at least partially, the destructive impact inherent to any excavation, making the process reversible through the 3D documentation of each context. It also grants greater flexibility as it allows context recording at times that are not necessarily constrained by the practicalities of the archaeological investigation. For this purpose, tests were performed to verify the possibility of detecting the sequence of the layers in the excavations through rapid photogrammetry sessions and without the aid of a total station. The result was obtained by realigning and scaling the models of the layers on the basic model, previously measured and georeferenced, and modulating the density of the meshes according to the purpose of use (Figure 1). The accuracy of the overlays on interventions of the size tested (up to 50 square metres) is fully comparable to that obtained through the use of markers and a total station.

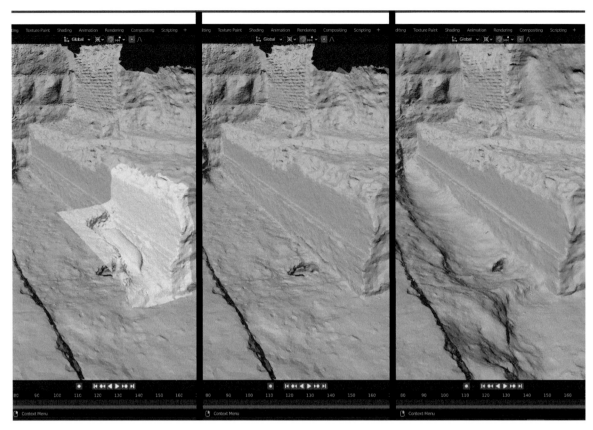

Figure 1. Overlay of archaeological layers (L. Bottiglieri, M. Solvi - 2021)

The model obtained, textured, inspectable and measurable with any software suitable for the manipulation of 3D models, was used as a support for the graphic investigation produced during the interpretation of the data, critically analysing plans, sections and elevations, highlighting the partitions, the stratigraphic units etc. and producing a technical survey of the excavation. Moreover, the execution of the entire production chain is also possible through the use of open source software such as: *QGIS* for the georeferencing of surveys, *meshroom* and others for photogrammetric analysis, *cloudcompare* and *meshlab* for the management of point clouds and meshes, *blender* for mesh management, 3D modelling, content creation, compositing and preparation of environments for virtual reality as well as real-time rendering engines such as *unreal engine* or *unity* for the use of content in virtual reality.

Integrated 3D model

In order to create a virtual model of the entire archaeological area of S Croce, to be used as a basis for the reconstruction of the various historical phases that have followed one another starting from the original morphology to finish at the most recent architectural phases, the available topographical and geomorphological data have been reorganized, inserting them on a georeferenced basis with the aim of analysing and comparing the floor plan , the alignments and reports of the evidence that have come to us.

We started by analysing the most ancient maps of Rome with topographic contour lines before the great morphological changes to Rome:

- Tavoletta F.150 IVSO – 'Roma', 1: 25,000 scale map, printed in 1877 by Istituto Geografico Militare (I.G.M.), with contour lines every 5 metres of altitude.

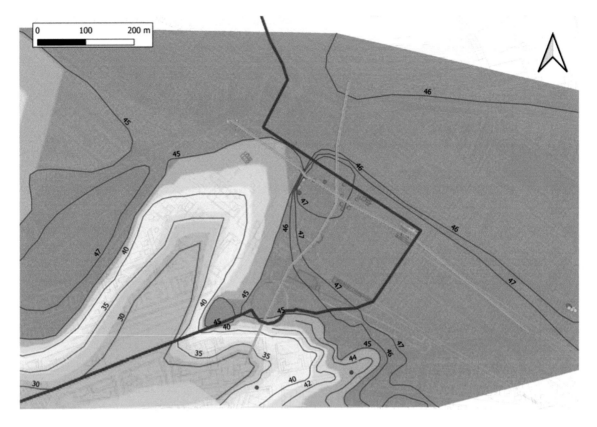

Figure 2. DTM of the original morphology of S. Croce in Gerusalemme Area (top of the geological substratum) obtained by archaeological and core drillings data and by literature and archive data. Isolines in meters a.s.l.; Green lines indicate the profiles of the geological sections. (C. Rosa – 2021)

- 'Piano Topografico di Roma e Suburbio', 1: 5,000 scale map, printed in 1908 by Istituto Geografico Militare (I.G.M.), with contour lines every 1 metre of altitude.

In these two maps, especially the one of 1908, a lot of backfilled areas are evident, especially in alluvial areas near the waterways, allowing us to filter this data to reconstruct the previous morphology.

The availability of numerous core-drilling stratigraphy data from different sources in the study area, together with archaeological data on ancient ground levels, allowed us to reconstruct, using QGIS open source software, a reliable ancient digital morphological model (Figure 2). The comparison of Figure 1 with some ancient maps of Rome (Bufalini, 1551; Del Re, 1557; Paciotti, 1557; Du Perac, 1577) confirms the quality of our reconstruction. This forms the basis for the subsequent work of the reconstruction of the evolution of the landscape.

The reconstruction of two geological sections, the first in a NW-SE direction and the second in a NNE-SSW direction (Figure 3), also allows us to verify the relationships between geology, morphology and choice of sites for the foundation of the different archaeological entities.

In detail, a single file organized by layers has included: the surveys and documentation of the excavations found in the archives, the images of historical maps, the aerial photos taken over time, the data of core drilling investigations so as to preserve all the available data, in order to allow to compare the evolution of the topography of the area and interpret the relationships between the finds. With the same principle, a three-dimensional model was created to integrate, in addition

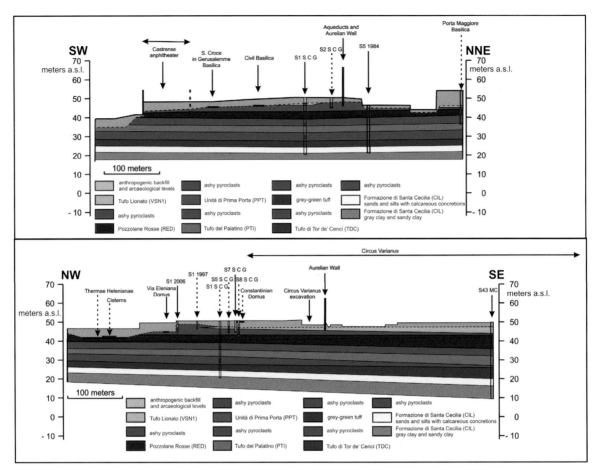

Figure 3. Geological cross-sections carried out by core drillings and archaeological data. (C. Rosa – 2021)

to two-dimensional information, also altimetric and 3D data, and to compare the morphological evolution of the area over time, the photogrammetric surveys and the reconstructions of the buildings.

The advantages of a 3D environment are many, starting with the opportunity of extracting data such as measurable floor plans elevations and ending with the possibility of studying their contents through platforms that allow the use of the models also in an exploratory way. The wide-ranging aim may be the insertion of data in an environment predisposed to fruition through virtual reality tools that, through the use of content streaming technologies, give the possibility to inspect and measure models, explore reconstructions, follow the evolution of studies and excavations in real time and remotely, with total immersion.

Reconstructions

As part of the realization of the project, reconstructions of the buildings of the Civil Basilica (so-called temple of Venus and Cupid) and of the domus along the aqueduct were carried out (fig.4). In accordance with the Seville Charter, a rigorous analysis methodology was held that would allow to select the data with objective certainty, highlighting instead the reconstructive integrations based on hypotheses

The Civil Basilica (so-called Temple of Venus and Cupid)

Following the restoration and excavation of 2015, a 3D reconstruction of the temple building was carried out by means of 3D modelling software, with scientific advice from Donato Colli and

Figure 4. The Civil Basilica (so-called temple of Venus and Cupid) and the domus along the aqueduct (L. Bottiglieri, D. Colli, S. Palladino, M. Solvi - 2021)

Sergio Palladino. Excavations in the apse discovered the floor screed, by which it was possible to reconstruct the kind of decoration that covered the floor with *opus sectile.* The internal decoration of the Basilica should have been characterized by a stylistic dichotomy: the big aula was created following the traditional Tetrarchic style with presence of columns and architraves along the long walls, creating chiaroscuro effects marked by niches with recesses and protrusion. On the other hand, the style inside the apse is different : according to the new Constantinian aesthetics, the walls and the ceiling had to be smooth to favour the diffusion of light and create an effect of architectural lightness. While the big hall had to be less illuminated, at the same time a maximum concentration of light in the apse would have underlined the importance of the location of the imperial throne.

As part of a project to enhance the heritage of the archaeological area, an archeo360 smartphone application has been created to provide visitors with a reconstructive hypothesis of the interior of the temple hall. The application works by reading spherical images that are uploaded by scanning a QR code and then processed using the phone's motion sensors. The spherical image is centred on the observer's point of view, i.e., on the spot where a pole with the code is fixed, and allows, by rotating the phone, to see through the screen the reconstructive hypothesis superimposed on the current building.

The domus by the aqueduct

The work of the reconstruction of the *domus* starts from a different principle because there are limited possibilities for the public to access the area and view its contents. The structure and size of the domus have led us to follow a path based on the production of content suitable for a representation in virtual or augmented reality. The excavation campaign currently underway is bringing to light new evidence mainly related to the context of the domus and to the close relationship with the adjacent structures of the Aurelian Walls and of the *Aqua Claudia*; at the end of the excavation campaign, the reconstruction will be completed on the basis of the new data that emerge and will be inserted in a space suitable for use in virtual reality, which allows a virtual visit to the domus in the various structural phases of the complex, highlighting the architectural and functional relationships with the aqueduct and the walls. Moreover, the digital 3D model will highlight which part of the visualization was based on the analytical study of structures and which one was otherwise based on comparisons with structures sharing similar function, design and chronology in Rome and beyond.

References

Barbera, M. 2010. Il comprensorio di S. Croce in Gerusalemme, novità topografiche ed archeologiche, *Bullettino della Commissione Archeologica Comunale di Roma* 111: 97-110.

Barbera, M. 2012. Aspetti topografici ed archeologici dell'area di Santa Croce nell'antichità, in R. Cassanelli, E. Stolfi (eds) *Gerusalemme a Roma. La basilica di Santa Croce e le reliquie della passione*: 1-11. Milan: Jaca Books.

Barbera, M. 2013. (ed.) *Costantino 313 d.C.* Milan: Electa.

Borgia, E., Colli, D., Palladino, S., Paterna, C. 2008. Horti Spei Veteris e Palatium Sessorianum: nuove acquisizioni da interventi 1996-2008 in *Fold&r - Fasti on Line Documents & Research* 124-125, republished in H. Di Giuseppe, E. Fentress (eds) 2014. *Roma. Scavi archeologici e scoperte degli ultimi 10 anni*: 101-142. Rome: Scienze e Lettere.

Borgia, E., Colli, D. 1998. Roma, S. Croce in Gerusalemme. Nuove acquisizioni topografiche. I Settembre-ottobre 1998. *Bullettino della Commissione Archeologica Comunale di Roma* 99: 243-246.

Bottiglieri, L., Colli, D., Palladino, S. 2016. Il comprensorio archeologico di Santa Croce in Gerusalemme a Roma: nuovi interventi di riqualificazione e recenti scoperte (2013-2014). *Bollettino di Archeologia On Line* 7(1-2): 133-144 .

Bottiglieri, L., D'Armini, T. 2020. I mosaici delle *domus* nel comprensorio archeologico di Santa Croce in Gerusalemme. *AISCOM XXVI*: 57-65.

Colli, D. 1995. Roma, Palazzo Sessoriano: nuove acquisizioni da disegni inediti di Edoardo Gatti. *Journal of Ancient Topography - Rivista di Topografia Antica* 5: 199-210.

Colli, D. 1996. Il palazzo Sessoriano nell'area archeologica di S. Croce in Gerusalemme: ultima sede imperiale a Roma. *MEFRA* 108: 771-815.

Colli, D. 2020. Costantino, il *Sol Invictus* e il palazzo Sessoriano. Spunti, dati e considerazioni per una ricostruzione della residenza imperiale. *Journal of Ancient Topography - Rivista di Topografia Antica* 30: 255-296.

From interpretation to 'provocation' and back again: Rome Transformed SCIEDOC and the Ospedale di San Giovanni in Laterano

I.P. Haynes[1], T. Ravasi[1], I. Peverett[2], M. Grellert[3], M. Simpson[4]

[1] Newcastle University, School of History, Classics and Archaeology (UK)
[2] New Visions Heritage Ltd (UK)
[3] Technische Universität Darmstadt (Germany)
[4] Newcastle University, Research Software Engineering Team (UK)
Ian Haynes - ian.haynes@ncl.ac.uk
Thea Ravasi - thea.ravasi@ncl.ac.uk

Integral to the interdisciplinary dialogue that drives the interpretation of the complex urban deposits encountered by the Rome Transformed (ROMETRANS) project is the notion of 'provocation'. A range of 3D visualisations are developed using architectural and landscape modelling software to make the team's interpretative work more tangible, accessible, and contestable. This approach has the merit of forcing team members to address issues that might be omitted from 2D plans and written descriptions. ROMETRANS works on the principle that visualizations developed by the team do not necessarily constitute a finished product but rather a 'provocation' to further engage team members, and the wider research community, in detailed discussion. To address the goals of the London Charter (2006) and Seville Principles (2011), the project supports these provocations with ROMETRANS SCIEDOC (https://rometrans.ncl.ac.uk/rtsciedoc/), a user-friendly system that allows all interested parties to explore and engage with the source materials used to justify each element, facilitating ongoing, open dialogue.

Recent years have seen several excellent responses to the requirements laid out by the European Network of Excellence in Open Cultural Heritage, and indeed to the London Charter and Seville Principles. We note with appreciation, for example, the Extended Matrix (Demetrescu *et al.* 2016; Demetrescu *et al.* this volume), the Cultural Heritage Abstract Reference Model (Gonzalez, Perez *et al.* 2012) and the creation of the Cultural Heritage Markup Language (Kuroczyński *et al.* 2016). Colleagues at Technische Universität Darmstadt developed the original SCIEDOC as an online tool to address the same imperatives (Pfarr-Harfst, Grellert 2016; Grellert *et al.* 2018). The system appealed to the ROMETRANS team from the outset, because in addition to responding to three fundamental principles of digital modelling of historical buildings: transparency, validity, and long-term availability, it also placed a particular emphasis on accessibility. There is no need to learn a new programming language, or to have a detailed knowledge of a given programme before starting to use the system.

In its original format, SCIEDOC is a repository for information and decision recording on projects focussing on individual buildings and working with relatively small teams. The scale of the ERC funded Rome Transformed project (Haynes *et al.* 2020), covering 68 hectares of research, with 23 archaeological complexes and two extended monumental structures such as portions of the Claudio-Neronian Aqueduct and a stretch of the Aurelian Wall, required a digital tool that could cope with scale and complexity. Between 2019 and 2021, a team of archaeologists and software engineers at Newcastle, with the constant insight and support of colleagues at Darmstadt, adapted SCIEDOC to the needs of Rome Transformed to develop ROMETRANS SCIEDOC, a tool that embeds the notion of visualizations as provocations introduced by the project and that reflects the complexity of the research area covered by the project, managing visualizations of buildings and portions of buildings alongside those large urban and natural areas. The ROMETRANS SCIEDOC website offers

Figure 1. Layout of the ROMETRANS SCIEDOC provocation page.

an introductory home page, a page dedicated to the provocations, a map of the area outlining where all the provocations are located and further pages for information and help. The core of the website lies in the provocation section (Figure 1), the space where provocations produced by the team are presented alongside the arguments underpinning them. This section provides further information on the complexities of the structure/area's biography, the methodologies used to recover data and the particular constraints on interpretation. Each provocation is also introduced by a summary description, which provides context, a map, and a link to all the supporting archaeological data that underpins the provocation. Provocations are searchable through filters and tags. The bottom portion of the provocation section offers the space for feedback and discussion.

To reflect better, for example, the complexity of a building's history from its initial design through its life and disuse and to allow maximum flexibility, ROMETRANS SCIEDOC has shifted the focus of the tool from the building itself to the provocation. Within the same archaeological complex, distinct spaces could belong to one property at one time and become part of separate properties at another: their visualizations are discussed as distinct provocations, but their original integrity is not lost, thanks to the use of tags that allow a quick search through all the provocations produced for the same architectural or archaeological complex. Tags also allow viewers to search ROMETRANS SCIEDOC, to focus for example on specific buildings or on topics they have an expertise on. Surviving evidence often permits more than one interpretation. Where this is the case, the argument section allows for the specific rationale to be discussed in support of each aspect of the provocation.

ROMETRANS SCIEDOC introduces a further step into the development of visualizations, underpinning our notion of provocation. Through the section dedicated to comments, it offers a dynamic online environment where provocations are used as visual props to encourage scholarly debate on every aspect of the visualization. Researchers can search through the visualizations using a map of the research area or can narrow down their search using dedicated tags should they wish, for example, to provide their feedback on a point of detail. The aim is to encourage and facilitate the engagement of specialists, allowing colleagues to be as specific as they wish, so that, for example, colleagues can target points of information for any element of the provocation. Visitors to the site can scrutinize the arguments underpinning each provocation and can provide written feedback or engage in a discussion that will be used to develop a new provocation, or Iteration. The comments section thus provides an additional space where the arguments underpinning the development of a new provocation are recorded and made available for debate. Importantly, this feature is also designed so that contributor's observations are themselves recorded, referenceable and acknowledged.

During the development of ROMETRANS SCIEDOC the project has made use of the provocations the ROMETRANS team has developed of the monumental fountain and of the portico of an elite residence discovered under Corsia Mazzoni in the Ospedale di San Giovanni in Laterano in Rome. These elements were excavated in a series of interventions, one sometimes before 1944 (Colini 1944) and the second in 1972 (Scrinari 1995), but surviving records are hard to interpret and vital source material is difficult to access, accordingly, and in keeping with the aims and methods of ROMETRANS, project members undertook a full survey/reappraisal of the surviving remains, published reports and unpublished source materials. The structures are part of the same archaeological complex and likely belonged to the same property: a *domus* dating to the Imperial period that underwent successive transformations, reaching its fully developed form during the 4th century C.E. when a large portico with a fountain at the centre and lavish floor and wall marble veneers were added. Despite being part of the same structural complex, the nymphaeum and the portico were never in use at the same time: the nymphaeum was partially dismantled when the portico of the residence was constructed. These two elements of the complex offer fine case studies to demonstrate the workings of ROMETRANS SCIEDOC.

After acquiring 3D data and getting the archival information about the historical excavations carried out at the site, the team carried out the structural analysis of all the surviving structures (Haynes *et al.* 2019; Ravasi *et al.* 2020a; 2020b). We then assessed the evidence and developed the phasing and interpretation of all the surviving structures: this is being made available as a preliminary report, including a summary of the main phases, a database of all the identified structural actions and the phase plans, from the Rome Transformed website (https://research. ncl.ac.uk/rometrans/). Finally, colleagues developed a series of provocations that have been made available through ROMETRANS SCIEDOC.

Figure 2. Provocation of the monumental nymphaeum discovered under Corsia Mazzoni (in the Ospedale di San Giovanni in Laterano, Rome. 1: variation with a semi-circular plan and half dome cover; 2: variation with a quarter of a circle plan and attic; 2: variation with a quarter of a circle plan without attic (Ravasi *et al.* 2020b)

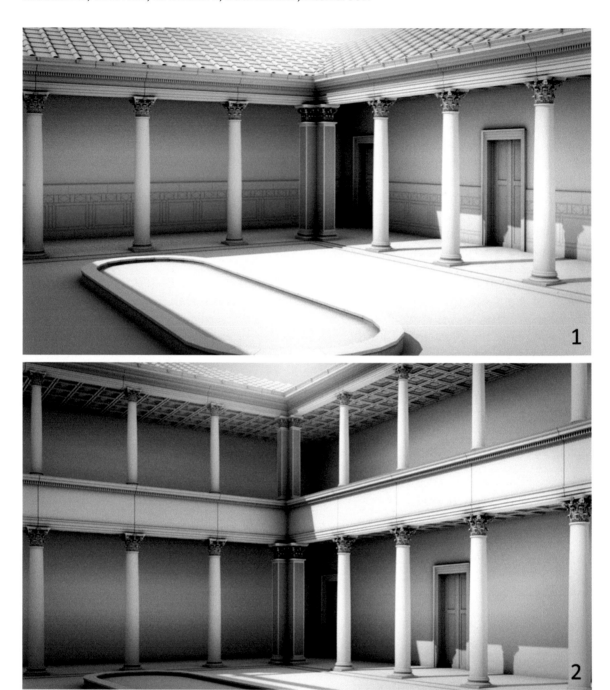

Figure 3. Provocation of the Late Antique portico of the domus discovered under Corsia Mazzoni in the Ospedale di San Giovanni in Laterano, Rome. 1: variation with one storey portico; 2: variation with a two storeys portico (Ravasi, Peverett 2019)

The remains of the nymphaeum are limited to the central portion of the structure, allowing for the identification of a series of alternate semi-circular and rectangular niches and of a waterspout. The bottom part of the structure, below the niches, has never been excavated, while the upper part was razed when the structures of the Ospedale were built. From a structural point of view, it is possible that the remains once belonged either to one semi-circular structure, possibly covered by a semi-dome, or to a quarter of a circle uncovered structure, either with or without an attic on top (Figure 2). The lack of information about the bottom part of the structure means that we are not sure whether the fountain featured a water basin at the bottom or a *stibadion*. The limited

Figure 4. Provocation of the Late Antique portico of the domus discovered under Corsia Mazzoni in the Ospedale di San Giovanni in Laterano, Rome (Ravasi, Peverett 2019)

surviving structures allow for the development of three variations of the same provocation, as many key elements are missing: a semi-circular fountain with a water basin at the bottom and a quarter of a circle fountain with a curved water basin at the bottom, one with an attic on top and one without it. The portico offers some degrees of structural variation in its northern and southern sides and the signs of its original marble decoration: this has allowed us to develop two main views of the same portico, one of the northern colonnaded side, with tall granite columns, and one of the southern arcaded side, with smaller columns placed on top of stone pillars and a low masonry parapet. Each view is also shown with its variants, for example, with a portico of one storey or with two.

Making project data available through ROMETRANS SCIEDOC has the value of forcing team members to justify their reasoning and to address questions of interpretation that might otherwise be omitted in textual descriptions or 2D images. It also requires collaboration between different elements of the project team, bringing together specialists in sites and materials, with colleagues who, while experts in the modelling of built and natural environments will not have the same site-specific expertise. Some provocations are presented as unrendered models (Figure 3), others as photorealistic representations (Figure 4). The use of lifelike three-dimensional visualizations in archaeology has been long debated: on one side are those who praise the role of photorealistic representations of the past as a means to foster debate and develop greater engagement with the public (Chng 2009; Pletinckx 2013; Rua, Alvito 2011; Sims 1997). On the other side, scholars have also emphasised the risk that lifelike representations can provide a false impression of complete knowledge, particularly when there is limited surviving archaeological evidence (Brusaporci 2016; Roussou and Drettakis 2004). We believe that the adoption of the term provocation and the development of more than one provocation for the same structure helps avoid such a risk. We actively seek reactions, rather than uncritical acceptance. Furthermore, we are privileged to have several structures in the ROMETRANS study area where sufficient remains of decorative elements survive for photorealistic provocations to be usefully and convincingly advanced. Developing an understanding of many of these buildings requires that both their art and architecture must be addressed in colour. The reader/viewer is nevertheless always actively encouraged to consider each visualization as a critical interpretation that is bounded by elements of uncertainty. ROMETRANS SCIEDOC allows us to make those uncertainties and limitations explicit. The inclusion of the full body of archaeological evidence underpinning each provocation helps avoid any potential misunderstanding of its nature and scope. Finally, photorealistic interpretations have a way or provoking responses especially if the viewer believes the interpretation is incorrect, harnessing a power to yield a criticism that can be used to advance research. In the context of a discussion, such as that offered in our feedback section in the provocation area of ROMETRANS SCIEDOC.

Such an approach has clear benefits for truly open data, but we hope to go beyond this, to foster the active participation of a still wider body of contributors to the research process, we hope that the provocative aspect of ROMETRANS SCIEDOC will enable this. To further advance this process, we aim to make ROMETRANS SCIEDOC fully referenceable, to ensure that discussion around provocations fully enters to the scholarly debate, and that contributors to that debate receive appropriate recognition and acknowledgement. An important facet of this will be to keep under review the degree to which colleagues engage with the platform, and study of what most encourages/discourages its use. The SCIEDOC platform has developed a tool that is simple to use, economic to implement, and replicable for the fulfilment of the London Charter and of the Seville Principles. Adaptation to the needs of ROMETRANS and the introduction of the concept of provocation will hopefully demonstrate the potential of such a tool for the future development of a standard practice for archaeological visualization and of platforms for scholarly debate on the visual outcomes of structural analysis and archaeological research. Embedding the London Charter and the Seville Principles will become a standard practice only if three key factors can be ensured: online databases provide tools sufficiently robust to support different kinds of visualizations, open access tools ensure that research groups can readily adapt such databases to their specific research needs, and a user-friendly interface is available that is not reliant on pre-existing technical knowledge or additional funding to operate.

Acknowledgments

Work on archaeology of the Ospedale began as part of the Newcastle/Florence SGL2 Project, funded through the British School at Rome (BSR), from a donation by a generous friend of the School. We would particularly like to thank colleagues at the Ospedale and Simona Morretta of the *Soprintendenza Speciale Archeologia Belle Arti e Paesaggio di Roma* for their generous support and advice. SCIEDOC was first developed by colleagues at Technische Universität Darmstadt. The development of ROMETRANS SCIEDOC forms part of the 'Rome Transformed' project has received funding from the European Research Council (ERC) under the European Union's Horizon 2020 research and innovation programme (grant agreement no. 835271).

References

Brusaporci, S. 2016. The importance of being honest: Issues of transparency in digital, in A. Ippolito (ed.) *Handbook of research on emerging technologies for architectural and archaeological heritage*: 66-93. Hershey: IGI Global.

Chng, E. 2009. Experiential archaeology: Is virtual time travel possible? *Journal of Cultural Heritage* 10 (4): 458–470.

Colini, A.M. 1944. *Storia e topografia del Celio nell'antichità, con rilievi, piante e ricostruzioni, di Italo Gismondi*. Rome: Tipografia Poliglotta Vaticana.

Demetrescu, E., Ferdani, D., Dell'Unto, N., Leander Touati, A. M., Lindgren S., 2016. Reconstructing the original splendour of the House of Caecilius Iucundus. A complete methodology for virtual archaeology aimed at digital exhibition. *SCIRES-IT* 6 (1): 51–66.

Gonzalez-Perez, C., Martín-Rodilla, P., Parcero-Oubiña, C., Fábrega-Álvarez, P. Güimil-Fariña, A. 2012. Extending an abstract reference model for transdisciplinary work in cultural heritage, in J. M. Dodero, M. Palomo-Duarte, P. Karampiperis (eds) *Metadata and Semantics Research. MTSR*

2012. Communications in Computer and Information Science, vol. 343: 190-201. Berlin, Heidelberg: Springer.

Haynes, I. P., Liverani, P., Ravasi, T., Kay, S., Peverett, I. 2019. The Lateran Project: Interim Report for the 2018–19 Season (Rome). *Papers of the British School at Rome* 87: 318-322.

Haynes, I. P., Liverani, P., Kay, S., Piro, S., Ravasi, T., Carboni, F. 2020. Rome Transformed: researching the eastern Caelian C1-C8 C.E. (Rome). *Papers of the British School at Rome* 88: 354-357.

Kuroczyński, P., Hauck, O., Dworak, D. 2016. 3D Models on triple paths – new pathways for documenting and visualizing virtual reconstructions, in S. Münster, M. Pfarr-Harfst, P. Kuroczyński, M. Ioannides (eds) *3D Research Challenges in Cultural Heritage II: How to Manage Data and Knowledge Related to Interpretative Digital 3D Reconstructions of Cultural Heritage*: 149-172. Cham: Springer International Publishing.

London Charter. 2006. Retrieved May 10, 2021 from http://www.londoncharter.org

Grellert, M., Apollonio, F. I., Martens, B. and Nußbaum, N. 2018. Working experiences with the Reconstruction Argumentation Method (RAM) - Scientific Documentation for Virtual Reconstruction, in *Proceedings of the 23rd International Conference on Cultural Heritage and New Technologies*. CHNT 23, 2018 (Vienna 2019). Retrieved May 10, 2021 from http://www.chnt.at/proceedings-chnt-23/

Pfarr-Harfst, M., Grellert, M. 2016. The Reconstruction – Argumentation Method: Proposal for a Minimum Standards of Documentation in the Context of Virtual Reconstructions, in M. Ioannides, E. Fink, R. Brumana, P. Patias, A. Doulamis, J. Martins (eds) *Digital Heritage. Progress in Cultural Heritage: Documentation, Preservation, and Protection*: 395-50. Heidelberg/Berlin: Springer.

Pletinckx, D. 2013. Archaeology and monuments in 3D in Europeana, in J. M. Dirk Callebaut, J. Mařík, J. Maříková-Kubková (eds) *Heritage reinvents Europe: Proceedings of the International Conference* (EAC Occasional Paper No. 7): 171-179. Namur, Belgium: Europae Archaeologiae Consilium (EAC).

Ravasi, T., Liverani, P., Haynes, I. P., Kay, S. 2020a. San Giovanni in Laterano 2 Project (SGL2). *Papers of the British School at Rome* 88: 350-354.

Ravasi, T., Haynes, I. P., Peverett, I. 2020b. The nymphaeum of Corsia Mazzoni: from archaeological investigation to 3D visualisation, in P. Liverani, C. Martini (eds) *Rileggere il Laterano antico. Il rilievo 3D dell'Ospedale di S. Giovanni, Workshops proceedings (Rome, 29th november 2018)*: 65-91. Florence: Insegna del Giglio.

Roussou, M., Drettakis G. 2004. Photorealism and non-photorealism in virtual heritage representation, in D. Arnold, F. Niccolucci, A. Chalmers (eds) *VAST 2003: Proceedings of the 4th International Symposium on Virtual Reality, Archaeology and Intelligent Cultural Heritage*: 51-60. Aire-la-Ville: Eurographics.

Rua, H., Alvito, P. 2011. Living the past: 3D models, virtual reality and game engines as tools for supporting archaeology and the reconstruction of cultural heritage: The case-study of the Roman villa of Casal de Freiria. *Journal of Archaeological Science* 38 (12): 3296–3308.

Santa Maria Scrinari, V. 1995. *Il Laterano imperiale. II. Dagli Horti Domitiae alla Cappella Cristiana*, Rome: Pontificio Istituto di Archeologia Cristiana.

Seville Principles. 2011. Retrieved May 10, 2021 from http://sevilleprinciples.com

Rome Transformed: a multiple method geophysical approach for the urban investigations of the East Caelian

S. Kay[1], E. Pomar[1], G. Morelli[2]

[1] The British School at Rome (Italy)
[2] Geostudi Astier Srl
Stephen Kay - s.kay@bsrome

Introduction

The Rome Transformed Project aims to improve our understanding of Rome and its place in cultural change across the Mediterranean world by mapping political, military and religious changes to the Eastern Caelian from the first to eighth centuries AD (Haynes *et al.*, 2020). The project is documenting the buildings and infrastructure that drove these changes, using a range of non-invasive techniques, with the aim of producing appropriately contextualised academically robust visualisations. A central component of the recording strategy is the use of geophysical prospection techniques suitable for deep stratigraphic urban investigations.

Urban survey challenges

The study area of the Rome Transformed project covers approximately 13.7km^2 of the eastern Caelian hill in central Rome. The densely populated area presents numerous challenges for geophysical prospection, ranging from urban infrastructure (tramlines, bus routes, electromagnetic disturbance), heavy traffic and modern services. The latter of these poses significant issues, as whilst many newer services are mapped, older or disused pipelines often do not appear in modern cartography.

A further challenge is presented by the deep urban stratigraphy. Excavations beneath the Archbasilica of St. John Lateran, which covers an area of 5,000m^2, have recorded palatial properties at a depth of circa 9m from the modern ground level. These buildings were destroyed and built over for the later Castra Nova, which itself was built over again for the Constantinian Basilica. The depth of stratigraphy, as well numerous complex occupation phases, also presents difficultly in both recording these structures with geophysics as well as the interpretation of the results. Moving to the east of the basilica, a similar overburden and build-up of the ground level occurs inside the Aurelian Walls, next to Viale Carlo Felice. Historical cartography of the area reveals topographically how much the area has changed, with the small hill of Monte Cipollaro to the west of the Castrense amphitheatre having been removed.

Methodology

The rich archival sources for central Rome, together with the detailed information in the SITAR database of the Soprintendenza Speciale Archeologia Belle Arti e Paesaggio di Roma (see Serlorenzi 2020 for a recent overview of the Archeositar project) including rescue excavations and environmental cores, allowed a detailed assessment to be made for each study area ahead of the geophysical prospection (Figure 1). The project chose to combine Electrical Resistivity Tomography (ERT) and Ground-Penetrating Radar (GPR), two methods suitable for urban investigations. In the study areas of the Circo Variano, Santa Croce in Gerusalemme and San Giovanni in Laterano it was possible to combine the two techniques, providing two comparative datasets recording different values. Furthermore, whilst the GPR in some areas, in particular at San Giovanni in Laterano, had a minimum amount of penetration, the ERT survey was able to support the assessment with a

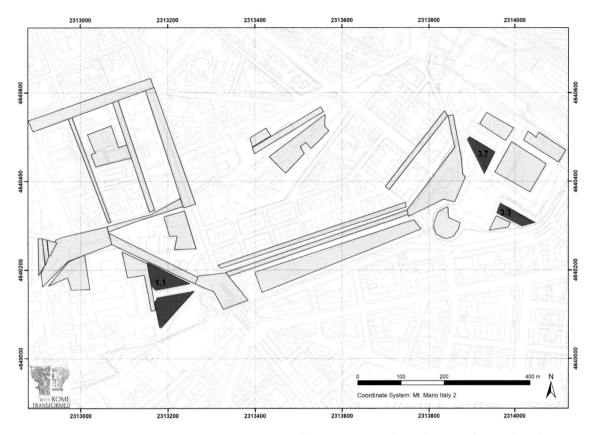

Figure 1. The geophysical prospection survey areas of the Rome Transformed project (in grey) with areas discussed in text highlighted in red.

greater degree of depth penetration (Piro *et al.*, 2020). The future analysis with the programmed environmental cores, the location of which will be partly dictated by the prospection results, will allow a better understanding of the subsurface through non-intrusive methods.

The separate surveys were recorded by GPS (Leica GS 18T) in Monte Mario /Italy Zone 2 (EPSG:3004), the same system adopted by SITAR, to precisely locate the geophysical surveys with the other datasets within the project GIS.

Study areas

The Rome Transformed project has applied both ERT and GPR at three sites within the study area. Elsewhere, the remaining 65 hectares will be investigated using GPR as the target areas are principally asphalt roads and pavements. The use of a multichannel system (IDS Stream Up) will allow for a rapid assessment with minimum disruption to traffic, an important criterion when working in a densely populated area.

Piazza San Giovanni in Laterano - Immediately to the east of the Archbasilica of San Giovanni in Laterano is an open piazza (Figure 1: Area 1.1), with a mixture of ground surfaces including grass, asphalt and *sanpietrino* cobbles. Excavations in the road have previously been conducted along the northern edge of the study area (between the basilica and the Scala Santa) but no further work appears to have taken place across the piazza. The area was previously investigated by GPR using a 400MHz antenna with traces of substantial structures recorded in the north-western area of the piazza, alongside Palazzo Lateranense (Piro *et al.*, 2020). The new GPR surveys undertaken by CNR as part of the Rome Transformed Project resurveyed the area with a 70MHz antenna as well as extending the survey to the southern part of the piazza. The results indicated a much more

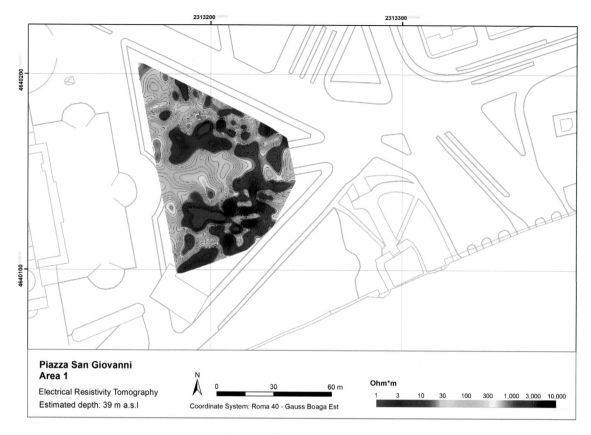

Figure 2. San Giovanni in Laterano (Area 1.1). ERT survey results at an estimated depth of 39m asl.

disturbed area to the south with difficultly in penetration beyond a depth of 2.5m. The full area was therefore investigated using ERT in order to assess the deeper stratigraphy (Figure 2). At a greater depth (circa 39m asl) a series of high resistance features were recorded, above all in the eastern part of the survey. The anomalies have a regular series of alignments, which correspond to the features recorded closer to the palazzo. Therefore, the combined methodology has allowed a higher resolution of data to be collected in the upper metres which is complimented by the deeper survey of the ERT.

Circo Variano - At the eastern end of the Rome Transformed study area lies the Sessorian Palace complex, including the Amphitheatrum Castrense, the Circo Variano, the so-called 'Temple of Venus and Cupid' and a number of residential houses of the 3rd – 4th century AD built alongside the Aurelian walls (Figure 1: Area 3.3). The survey area of the circus lies immediately behind the Basilica of Santa Croce in Gerusalemme and covers an area of approximately 50m by 20m with an area of open excavations of the circus running along the northern edge of the survey area. The GPR survey, conducted using a 200MHz antenna with a traverse separation of 0.5m, revealed features in the upper levels (at approximately 0.8m) but beyond a depth of 1.5m few clear features were recorded (Figure 3). To assess the deeper levels the same area was also investigated with ERT with 6 connected profiles laid out at a distance of 6m and a probe separation of 2m. An area of high resistance was recorded both alongside the circus, but also on a different alignment in the south-eastern part of the survey (Figure 3). The high resistance anomalies to the north clearly relate to unexcavated parts of the circus structure, whilst the feature to the southeast will be further invested through an environmental core to assess the composition of the feature.

Temple of Venus and Cupid - The area of the so-called 'Temple of Venus and Cupid' lies within the grounds of the Museo Storico della Fanteria in the north-western part of the Sessorian Palace

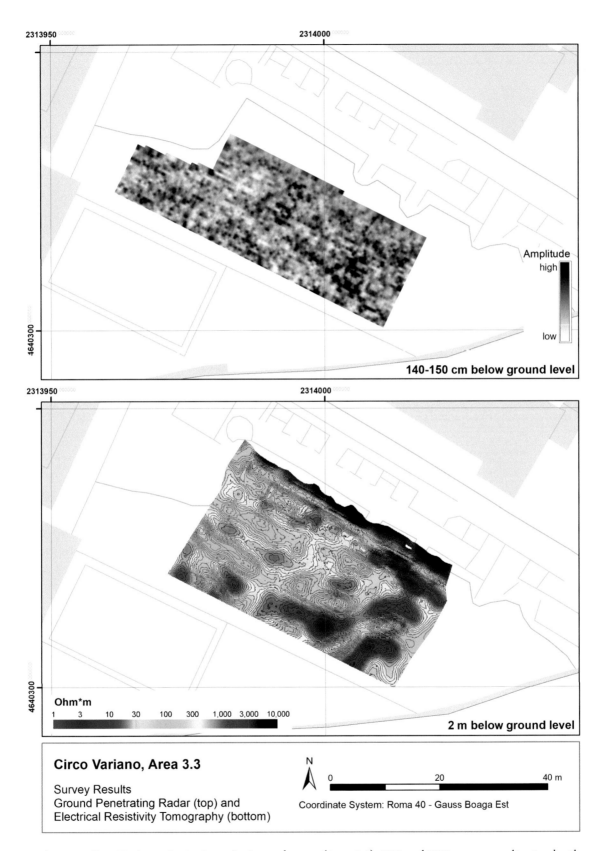

Figure 3. Circo Variano, Santa Croce in Gerusalemme (Area 3.3). GPR and ERT survey results at a depth between 1.5m and 2m below ground level.

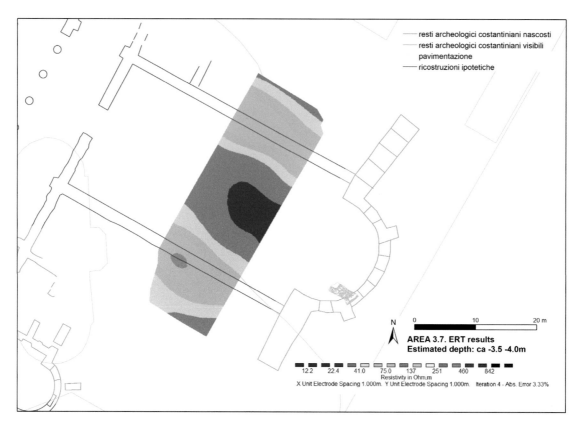

Figure 4. Temple of Venus and Cupid (Area 3.7). ERT survey results at an estimated depth of 3.5-4m below ground level.

complex (Figure 1: Area 3.7). The gardens within the museum grounds offered an opportunity to further explore the complex, parts of which have been investigated by the Soprintendenza Speciale Archeologia Belle Arti e Paesaggio di Roma. A number of different GPR antennas were used (400MHz, 200MHz and 70MHz) as initial profiles indicated a difficulty in penetration, potentially due to the high water content in the soil. An excavation in the apse of the structure in 2013-2014 (Bottiglieri *et al.*, 2016) recorded the floor preparation of the building at a height of 46.29m asl. The ground level in this area is circa 50m asl, a difference of 3.71m, beyond what was recorded with any clarity by the GPR survey. The same area was therefore investigated with ERT using a probe and traverse separation of 1m (Figure 4). The survey recorded two parallel areas of higher resistance, in correspondence with the hypothesised northern and southern walls of the basilica (Bottiglieri *et al.*, 2016). The depth of the features, and their subsequent continuation to a depth of approximately 6.5m corresponds with the floor surface of the basilica.

Conclusion

Geophysical prospection in the centre of an urban area of Rome presents numerous challenges requiring a careful strategy of assessment of the known archaeological record ahead of any survey. However, the communal parks, gardens and in particular the roads offer an opportunity to investigate large parts of the subsoil. The methodology used by the Rome Transformed project, combining low frequency GPR antennas with ERT and subsequent investigation with environmental coring provides a series of complementary datasets, rigorously recorded in their spatial context. When such information is then assessed with excavation and archival information, a better understanding can be achieved of the subsoil.

Acknowledgments

This project has received funding from the European Research Council (ERC) under the European Union's Horizon 2020 research and innovation programme (grant agreement no. 835271). The project is fortunate to have a wide consortium of partners in Rome who have both facilitated the research as well as share an immense knowledge of the study area: Soprintendenza Speciale Archeologia Belle Arti e Paesaggio di Roma; Comune di Roma – Sovrintendenza Capitolina ai Beni Cultural; Musei Vaticani. The project is also grateful to the British Embassy in Rome (Villa Wolkonsky), the Museo Storico della Fanteria and the Istituto Santa Maria for permitting access to their properties.

References

Bottiglieri, L., Colli, D., Palladino, S. 2016. Il comprensorio archeologico di Santa Croce in Gerusalemme a Roma: nuovi interventi di riqualificazione e recenti scoperte (2013-2014), *Bollettino di Archeologia On Line*, VII (1-2): 133-144.

Haynes, I. P., Liverani, P., Kay, S., Piro, S., Ravasi, T., Carboni, F. 2020. Rome Transformed: Researching the Eastern Caelian C1-C8 CE (Rome). *Papers of the British School at Rome* 88: 354-357.

Piro, S., Haynes, I. P., Liverani, P. and Zamuner, D., 2020. Ground-Penetrating Radar survey in the Saint John Lateran Basilica Complex. In L. Bosman, I. P. Haynes, P. Liverani (eds) *The Basilica of Saint John Lateran to 1600*. British School at Rome Studies: 52-70. Cambridge: Cambridge University Press.

Serlorenzi, M., 2020. Il SITAR a supporto di una ricerca condivisa: per una nuova pianta di Roma medio repubblicana, in A. D'Alessio, M. Serlorenzi, C. J. Smith, R. Volpe (eds) *Roma medio repubblicana. Dalla conquista di Veio alla battaglia di Zama*. Atti del convegno internazionale: 11-26. Rome, Edizioni Quasar.

Three coloniae and three municipia: non-invasive exploration of urban contexts in Roman Hispania

L. Lagóstena, J. A. Ruiz Gil, J. Pérez Marrero, P. Trapero, J. Catalán, I. Rondán-Sevilla, M. Ruiz Barroso.

Departamento de Historia, Geografía y Filosofía, Facultad de Filosofía y Letras, Unidad de Geodetección, Análisis y Georreferenciación del Patrimonio, Universidad de Cádiz (Spain)

José-Antonio Ruiz Gil - jantonio.ruiz@uca.es

Lázaro Lagóstena Barrios - lazaro.lagostena@uca.es

Jenny Pérez Marrero - jenny.perez@uca.es

Pedro Trapero Fernández - pedro.trapero@uca.es

Javier Catalán González - javier.catalan@uca.es

Isabel Rondán Sevilla - isabel.rondan@uca.es

Manuel Ruiz Barroso - manuel.ruiz@uca.es

Introduction

Since 2016 our team has been developing non-invasive research methodologies applied to the study of large archaeological areas corresponding to ancient cities in Roman *Hispania*. The Roman colonies of *Hasta Regia* (Mesas de Asta, Jerez de la Frontera), *Ilici* (La Alcudia, Elche) and *Libisosa* (Lezuza, Albacete), and the Latin municipalities of *Balsa* (Luz de Tavira, Faro), *Arva* (Alcolea del Río) and *Calduba* (La Perdiz, Arcos de la Frontera) located in the Roman provinces of *Baetica, Tarraconensis* and *Lusitania* have been used as case studies presented in this paper (Figure 1). The cities in this each differ, offering a wide variety in terms of their topography, geography, current uses, and states of archaeological intervention and conservation. Furthermore, research at each of them has been approached with due consideration to these particular conditions, while experimenting with the workflow that was considered more appropriate to develop the non-invasive research and ensure the best results. This paper discusses the different conditions and problems during the entire research process and the strategies applied, as well as the most relevant results in each case study as part of the conclusions from this experience. The main techniques and tools used during this work have been Unmanned Aerial Vehicles (UAVs), ground-penetrating radar (GPR), magnetometry, photogrammetry and terrestrial Lidar. Due to the number of sites studied, the period of the studies, the nature of the surfaces analysed, and the diversity of situations, this research experience is unique among other Spanish teams specialising in non-invasive techniques.

Geographic characterization of cases

Colonia Hasta Regia (Mesas de Asta, Jerez de la Frontera, Cádiz). This site is located in the Lower Guadalquivir, in a geological plateau formation, 78 meters above sea level and approximately 25 hectares *intra moenia*. The settlement's origins begin in the Final Tartesic Bronze Age and it was a key site for the territorial control and organization in the Lower Guadalquivir (*Baetis*). Punic influence can be seen in the evolution of the site, before the settlement's conversion in the Late Republican period into a colony in the Roman province of Baetica. The city continued to be inhabited until the Late Antiquity and Islamic Medieval periods. Its location was nearly insular, surrounded by marshes because of the Guadalquivir River estuary. Today the site is uninhabited and lies on farmland, cultivated with different crops including olive trees. Very few archaeological excavations have taken place here (Martín Arroyo 2018).

Colonia Iulia Ilici Augusta (La Alcudia, Elche, Alicante). The *Ilici* settlement is located on a small hillock, surrounded by cultivated fields in the meadows of the River Vinalopó. It lies 80 meters

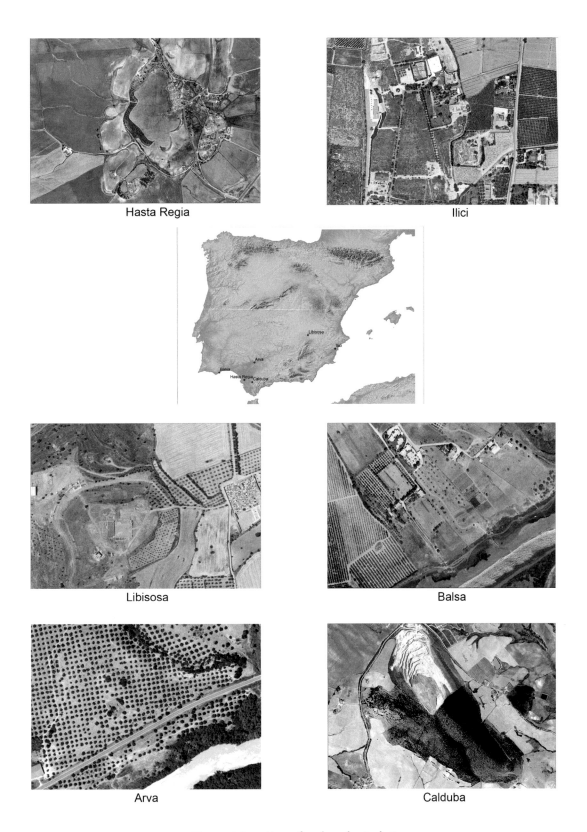

Figure 1. Location of archaeological sites

above sea level and covers a total surface of nine hectares. Following its Iberic origin, it was converted into a colony around 43-42 BC. As with the previous example, the occupation continued until the medieval period. The area within the city walls has a gradient profile, with small platforms terraced to be flat, characteristic of a traditional irrigation system. The site has been extensively investigated by archaeologists since the nineteenth century, with excavations in outstanding areas of the site (Tendero, Ronda, 2014).

Colonia Libisosa Foroaugustana (Lezuza, Albacete). Located near Lezuza on a hill between 950-1000 meters above sea level and covering approximately 30 hectares, *Libisosa* was an Iberic-Oretan *oppidum* in origin, transformed into Roman Colony and occupied into the medieval period. It has preserved some of its defences. The archaeological site is terraced, the central and highest terrace is occupied by the forum while the rest of the settlement extends down the surrounding hillsides. There has been an intense archaeological intervention at the site in recent decades (Uroz 2012).

Municipium Balsa (Luz de Tavira, Tavira, Faro). The archaeological remains of the ancient *Balsa* are near the coast, in the actual marshland of the Natural Park of Ría Formosa. The Roman *municipium* was founded during the Flavian dynasty in a pre-Roman community. The city could have reached a surface coverage of 47 hectares, being one of the biggest all over the Lusitanian province. In archaeological interventions, it has been documented ports, thermal buildings, canals and sewage systems, fish salting factories and burial grounds. In addition, it could have entertainment buildings, a theatre and a circus according to the epigraphic evidence, although these have not been located yet. The site is located on a slight slope towards the marshes, with a geological composition of consolidated calcareous sand and gravel. It is currently divided into several independent agricultural plots, with a variety of crops, and other residential properties. (Gil Mantas 1990).

Municipium Flavium Arvensis (Alcolea del Río, Sevilla). This settlement, of pre-Roman origin, became a *municipium* in the Flavian period. The *oppidum* was located on the banks of the Guadalquivir River, while the Roman city was located on a plateau between 50 and 62 meters above sea level. The urban nucleus lies on the highest part and covers about 4 hectares. There has been little archaeological intervention at the site, although the discovery of large thermal baths and amphora-producing potteries in the suburb is noteworthy (Remesal *et al.* 1987). The entire site is occupied by a traditional olive grove.

Municipium Calduba (Sierra Aznar, Arcos de la Frontera, Cádiz). There is little evidence of pre-Roman activity at this site, which reached its height during the 1st century AD. Thereafter it retained a residual population into the Late Roman and Andalusian periods. It is located in the basin of the Majaceite River, a tributary of the Guadalete river and occupies a mountainous, terraced area with steep slopes between 200 and 400 meters above sea level. The site is well known for its important hydraulic structures, but the recent archaeological interventions carried out in the last two decades have consisted mainly of a superficial cleaning of these structures, and not of further investigation. Although not fully confirmed, the Sierra Aznar site is most probably the municipality of *Calduba*, cited by Claudius Ptolemy (Geog. II. 3; Lagóstena 2015). The site is mostly covered by a natural woodland , with a predominance of wild olive and holm oak trees.

Techniques applied and results

Colonia Hasta Regia - The *Hasta Regia* study is a University of Cadíz research project. It began in 2016 with the use of drones, cartography, orthophotography and other precision survey methods. Thereafter, a phased GPR exploration was planned, starting from the south towards the north of the site, working with NE-SW passes adapted to the orientation of the known archaeological structures, the orographic conditions of the site, especially the slopes and the water run-off system that crosses it. Sixty per cent of the surface area (15 ha) has been surveyed with GPR. Exploration

Figure 2. Hasta Regia and Ilici cases

of the city necropolis with magnetometry and part of the unexplored site with geo-radar began in 2020, covering an area of 13 hectares (Figure 2.1). As a result, the GPR reveals information about the classical and post-classical urban periods but is less effective for the pre-Roman ones because we have no penetration capability and it is very likely that the structures detected are Romans. This is because the configuration of the soil is essentially clay, whose electrical conductivity is very high, making it difficult to see results at depth. On the other hand, magnetometry has made it possible to obtain a mapping of great interest for the knowledge of the extensive necropolis, where a micro-survey had been carried out in the 1990s (González *et al.* 1992). At present, work continues with the post-processing of the data obtained.

Colonia Iulia Ilici Augusta - The work in *Ilici* was carried out at the request of the University of Alicante. The objective was to obtain a GPR image of the subsoil of the city and to support the archaeological work of different research teams in different areas of the city. The exploration began in 2017 and was methodologically tailored to the plot configuration of the site, adapting the survey to the traditional irrigation terraces. This work covered approximately 3.3 hectares, most of the site had not been excavated or was not occupied by existing buildings (Figure 2.2). The result reveals information very relevant to the urban structure of the colony. Some building elements have also been documented that probably relate to the pre-Roman phase of the city, in the area where the famous bust of the 'Dama de Elche' was found. The soil conditions for exploration were very suitable (vegetation clearance, small slopes and good composition of the substrate).

Colonia Libisosa Foroaugustana - In the case of *Libisosa*, the singularity consists in the use of geo-radar in the plots of the city still unexcavated, obtaining information of great interest to complete the whole image of the city. The work forms part of projects led by the University of Alicante and Murcia. In addition, already excavated plots have been explored, particularly the city's forum, obtaining relevant information on previous phases of occupation and other unknown elements of interest, which are currently under study (Figure 3.1). Working conditions were not suitable, however, due to the existence of tree plantations and the steep slopes and terraces of the terrain. For the analysis of some of the explorations, we relied on aerial photogrammetry. Two GPR devices were used, one for large areas (IDS multi-channel Stream) and another single-channel, a two-frequency device for areas that are difficult to access (IDS Hi-mod). In addition to the producing results that have enhanced our understanding of the urban fabric of *Libisosa*, we make a study of cumulated images in a post-processing phase that substantially improved the visualisation of the archaeological structures.

Municipium Balsa - For the exploration of *Balsa*, carried out at the request of the University of Algarve, a 3D Radar multi-frequency geo-radar was used. The exploration was carried out on various agricultural plots, with different conditions in terms of topography, ground cover, crops and geology, giving diverse results. Previous GPR and magnetometer surveys have been carried out on the site, with interesting results. Work continues with multi-channel 200 Mhz GPR and magnetometer. The results illuminate important aspects of the urbanisation of the city, its urban structure, production centres (salting fish industries) and, possibly, public entertainment areas that had not previously been located (Figure 3.2). A sector of the necropolis was also successfully surveyed using multi-frequency equipment. The greatest difficulty lies in connecting the results recovered from the different plots, although future work will progress this.

Municipium Flavium Arvensis - The work in *Arva* has been promoted and financed by the Alcolea del Río Town Council. The objective is to ascertain the condition of the archaeological remains to support the study and heritage recovery of the latter. The greatest difficulty facing the survey team lies in the presence of an olive grove that covers the entire site, planted in a 10 x 10-metre framework. The work carried out with GPR (single and multi-channel) and a magnetometer has been adapted to these circumstances, with the support of aerial and terrestrial photogrammetry of the visible archaeological remains (Figure 4.1). We obtained information of great interest for the

Figure 3. Libisosa and Balsa cases

Figure 4. Arva and Calduba cases

knowledge of the town planning of the municipality and its most outstanding elements, as well as that of the artisanal suburban area along the banks of the river. The experience of exploration with large equipment in areas with wooded crops has also been unique and successful.

Municipium Calduba - The case of Sierra Aznar is one of the most complicated due to the complex orography and difficult conditions. In this case, documentation by aerial and terrestrial photogrammetry has been of great importance. In the first case, a large number of photogrammetric flights have been carried out, together with LiDAR analysis (national flights IGN) of the site that has also been used to obtain data on those elements. To study the abundant visible archaeological elements, we combine classical land photogrammetry with the use of a drone and terrestrial laser scanner. The first challenge of the site is to understand the original planning and organisation. The link between the archaeological remains and the use and management of water is a very relevant factor, as the municipality may have had a close relationship with water cults. The hillside on which the remains are located was constructed in the late Imperial period to create an important system of terraces, which developed over a difference in elevation of 200 metres in some 450 linear metres (Figure 4.2). Geophysical prospecting work will begin once the general shape of the site and the priority research agenda have been established. GPR equipment will be used, especially to search for irrigation and drainage systems, and the magnetometer will be used to locate the necropolis of the settlement.

Concluding remarks

Work carried out by our team over the last five years has been very intense. We have been confronted with a significant number of case studies, which present very varied casuistry, but all correspond to relevant urban settlements. Most of the settlements under discussion began in the protohistoric period, with their *floruit* in Antiquity, and practically all of them show occupation at least until the Islamic period. In some cases, these are settlements with complex stratigraphy, where traces of several buildings and occupational phases remain, and this presents a challenge when applying non-invasive research methods to study them. The diversity of conditions in terms of geography, orography, landscape and geology, has forced us to adapt the instruments and the exploration methodology to each specific case. We have used the various technical resources available for the best analysis of each case our work has also been framed by the objectives of the research projects that have required our collaboration.

In all cases the results are highly relevant, undoubtedly providing invaluable information in each case study, all of which will be published in detail later. In our opinion, the test we have carried out of various protocols for the application of the non-invasive methodology in the survey of large areas, in such diverse conditions, is also very outstanding.

References

Gil Mantas, V. 1990. As ciudades marítimas da Lusitania. J.-G. Gorges (ed.) *Les villes de Lusitanie Romaine*: 149-205. Paris: Éditions du Centre National de la Recherche Scientifique.

González Rodríguez, R., Barrionuevo Contreras, F., Aguilar Moya, L., Ruiz Mata, D. 1992. Prospección arqueológica superficial en el entorno de la marisma de Mesas (Jerez de la Frontera, Cádiz). *Anuario Arqueológico de Andalucía* II: 71-77.

Lagóstena Barrios, L. 2015. La obra hidráulica romana en la cuenca del Guadalete. *Río Guadalete*: 148-156. Sevilla: Consejería de Medio Ambiente y Ordenación del Territorio, Junta de Andalucía.

Martín-Arroyo Sánchez, D. J. 2018. *Colonización romana y territorio en Hispania. El caso de Hasta Regia*, Collecció Instrumenta 61. Barcelona: Universitat de Barcelona.

Remesal, J., Revilla, V. Carreras, C., Berni, P. 1987. Arva: prospecciones en un centro productor de ánforas Dresel 20 (Alcolea del Río, Sevilla). *Pyrenae* 28: 151-158.

Tendero, M., Ronda, A. M. 2014. La ciudad romana de Ilici (L'Alcudia de Elche, Alicante), in M. Olcina (ed). *Ciudades romanas valencianas*: 225-243. Alicante: Museo Arqueológico de Alicante - MARQ.

Uroz, J. 2012. La colonia romana de Libisosa y sus precedentes, in G. Carrasco (ed). *La ciudad romana en Castilla-La Mancha*: 87-130. Ciudad Real: Universidad de Castilla-La Mancha

The topography of Rome. An outlook for the future

P. Liverani

Universitá degli Studi di Firenze (Italy)
Paolo Liverani - paolo.liverani@unifi.it

The studies concerning the topography of Rome have a long tradition stemming from the Renaissance. The status of this field of research as a scientific and academic discipline dates back to the nineteenth century thanks to scholars like Heinrich Jordan, Christian Hülsen, Rodolfo Lanciani and – in the early twentieth century – Thomas Ashby. In the last generation this field underwent a dramatic renovation. On the one hand there have been great archaeological excavations such as the *Crypta Balbi* or the Imperial Fora, on the other there has been a more mature awareness of the importance of studying the urban phenomenon over the *long durée*, with equal attention to all historical periods.

The topography of a city – and the case of Rome is paradigmatic – is an invaluable source in its own. Its knowledge has important repercussions that go far beyond the reconstruction of the monumental fabric, to have an impact on economic, political, and religious history. Furthermore, it is evident that a quality leap is needed to integrate geographical, geomorphological, and environmental research. With this target in mind, it is clear the absolute need of a three-dimensional knowledge of the city, overcoming the traditional two-dimensional plans, which can neither document the superimposition of the phases, nor help to visualize the elevation of the buildings forming the armature of the city. Both these points are essential for an appropriate characterization of the various urban districts from a monumental and functional point of view. Furthermore, both in ancient and modern times the city underwent dramatic changes in its physical morphology: on the Caelian for example, I can mention the terracing for the construction of the *Castra Nova Equitum Singularium* and later the building of the Aurelian Walls that cut through existing small valleys and secondary rivers, causing a series of backfills in the following centuries. More recently, in some instances the existing difference in altitude has been softened, like in the Piazza Porta S. Giovanni or in the area to south of the basilica of S. Croce in Gerusalemme. As a result, the modern Caelian hill appears much flatter and more levelled in comparison with the ancient landscape.

A 3D approach to the topography of the city has some necessary implications. On one side the need of three-dimensional surveys of all the surviving structural evidence, with various methodologies, such as photogrammetry, laser scanning, structure from motion etc. On the other, the systematic employment of geognostic surveys for the subsoil, possibly integrated by drill core campaigns, in order to establish the depth and the thickness of the archaeological deposits. Finally, the development of one or more Digital Terrain Models corresponding to distinct transformations of the urban fabric of the city.

The first need – the 3D survey of the monuments – has its own methodological implications, but it is not specifically the focus of this conference. The second one, the knowledge of the buried city mainly through non-intrusive methodologies, is at the core of our meeting. The issue needs to be considered from two perspectives. The first, more obvious, is the technical aspect, and we have several esteemed colleagues in our conference, highly skilled in the field, willing to share with us many innovative and exciting experiences. The importance of this approach is self-evident. Regardless of costs, it is very difficult to have opportunities for extensive excavations in the densely occupied areas of the city and, when they take place, they generally come in the shape of rescue excavations linked to construction or to renovation works or to the setting up of public facilities. Generally, they do not stem from planned academic research. Therefore, a systematic exploration of unbuilt, open areas such as gardens, parks, squares, and streets can provide, with a little bit of luck, a great amount of information. Even considering the unavoidable presence of areas lacking any surviving evidence because of the recent history of the city, or of disturbances due to multiple reasons, the result of these surveys, especially when different approaches are integrated, can provide at least some general information on the layout of the district that can be useful to understand the layout of the urban fabric.

The second point is not strictly academic but connected with the positive repercussions on the urban planning and management, offering an invaluable tool to the authorities responsible for the protection of cultural heritage to respond in a quick way to the urban development programs, limiting to the essential the slower and more expensive excavations of preventive archaeology.

To digress for a moment to a more general consideration of the Italian situation, which at the moment is quite critical. After a period when the protection system of the cultural heritage was quite strong and applied with rigour, in the last years the political orientation moved in the opposite direction. The last reform of the Ministry of the Cultural Heritage had an extremely negative effect: now the fragmentation of the territorial offices – the Soprintendenze – the shortage of their funding and staff, the weakening of the tools and of the procedures to control the territory are evident especially when under pressure for natural disasters – like earthquakes – or large development plans on the territory. At the same time there are growing difficulties in the collaboration between the Ministry of the Cultural Heritage and the Ministry of the University. I must be extremely clear to avoid misunderstandings. The collaboration between university departments and territorial offices is as good as ever and in our case is simply ideal: we always encountered full understanding for our needs and prompt and friendly cooperation from all the colleagues of the Soprintendenza. On the other hand, a series of subsequent ministerial circulars imposed increasingly stringent conditions to archaeological research for universities. The last one had a very strong impact just on non-intrusive methodologies, raising strong criticism and protest from the universities and embarrassing the colleagues of the territorial offices. A subsequent circular changed some details, slightly improving some of the most indefensible parts of the new rules, but in general the situation remains very critical both in principle and in practice. I hope these restrictive provisions will be removed, but at the moment there is no positive sign in this direction.

But it is time to go back to the topography of ancient Rome. The more articulated questions and the higher standards required by contemporary research have a series of implications. The range of different fields of research, the technical improvement in the survey, the amount of data to consider and to organize require research groups made up of specialists from several disciplines and well-defined programs, in order to bring together all the evidence in an organic synthesis. Last but not least, this great bulk of data and results need to be shared with the scientific community, preserved in an efficient way so that it can be used in future researches, and transferred both to the general public and to the authority responsible for the planning and the management of the city.

The topography of ancient Rome has always had in his DNA a methodological connotation and an interdisciplinary vocation. Since the beginnings, this field of study needed to face with cartography, architectural survey, structural analysis, philological analysis of literary and epigraphic sources, iconography, numismatics, archival research in a long-term perspective. As we have seen, this multifaceted approach is pushed far beyond the limits of the disciplines that a single researcher can master. Even if a lot of specific research can and must still be carried out by single researchers with high specialization, more substantial and innovative results need the establishment of close-knit research groups and the elaboration of integrated research programs. The goal is to build a methodological and technical frame, which can be replicated and progressively improved. In this way it will be possible to compare the results, to provide more comprehensive synthesis, but also to obtain powerful tools enabling the authorities in charge for the protection of the cultural heritage and for the territorial development to respond in the most effective way to the urban evolution.

On the basis of these premises there are some consequences even for the more academic core of the discipline. The first task that the topographical study of Rome had to face was the reconstruction of the city plan, combining a rich but at the same time extremely fragmented literary and epigraphic evidence with the not less fragmented archaeological remains or with the traces and memories of the ancient monuments recorded during the centuries. After this unavoidable first step of the topographical research, the effort followed to reconstruct a narrative of the urban evolution, studying the social, political and economic functions of the various districts, exploring their religious and ideological connotations, the road network and its ceremonial use in the different periods.

At the same time there was a growing consciousness of the need to connect the classical and the medieval evolution of Rome in a single narration. In the past the two fields had a rigid separation only partially bridged by the specialist of the Early Christian archaeology.

What is now the task of the next generation of scholars in this field? Obviously, I do not have my crystal ball, nor do I pretend to outline the future of such a complex discipline. However, despite all, we can try to guess at least a couple of points where a progress is needed and is at hand.

First of all, in the future the need to document the city in a systematic way in its physical structure with a close attention to its morphology, placing the archaeological and monumental evidence in the three spatial dimension or, better, in the four dimensions (with time considered the fourth) will be more and more clear. Only in this way we will be able to advance beyond Lanciani's *Forma Urbis Romae*, which still after a century is the most comprehensive image of the city.

Secondly, we need to connect the urban research of the ancient periods to the later periods, until the modern times. We already address the *long durée* in excavations and in the study of the history of the buildings and of the urban evolution, but the tools for the study of the ancient urban planning are still limited in their potential. We still privilege specific moments and specific types of cities in the study of ancient urbanism: for example, the time of the foundation of Greek and Roman colonies, where we can recognize a clear plan, a project, an idea of city. But by no means can these notions pretend to represent the entirety of the ancient urban phenomenon and we still have very few tools to assess the urban transformation in later periods. Furthermore, if we compare books dealing with the ancient city with others concerning a medieval or modern one, it is nearly impossible to build a single story, because questions and methodologies differ completely, even if the cities, at least in part, are the same.

The difference of approach is mostly due to the fragmentary evidence the archaeologist has at their disposal for reconstructing the ancient urban fabric, compared to the rich cadastral documentation of the early modern cities. In the case of Rome, we have a little part of the cadastral map of the ancient city but, above all, we have an enormously rich body of evidence both from the archaeological excavations and the written sources. If correctly systematized, this evidence could give enough elements that would allow us to use, even if in a simplified way, some of the tools already well experimented in the studies of the urban geography of medieval and modern cities. I am referring to the so-called Urban Morphology: this type of approach could also be useful in some other instances, where we have extensive excavations like Ostia or in the study of the Roman cities of northern Africa.

In this way we could try to put an end to this meaningless division in urban studies, between the study of ancient and the modern times and develop a dialogue between the specialists of these two fields of research to bring together their different methods in a shared endeavour.

Ground-penetrating radar survey as the linchpin of a multidisciplinary approach to the study of two Roman cities in Lazio

A. Launaro[1], M. Millett[1], L. Verdonck[2], F. Vermeulen[2]

[1] Faculty of Classics, University of Cambridge, UK
[2] Department of Archaeology, Ghent University, Belgium
Alessandro Launaro al506@cam.ac.uk
Martin Millett - mjm62@cam.ac.uk
Lieven Verdonck - lieven.verdonck@ugent.be
Frank Vermeulen - frank.vermeulen@ugent.be

Introduction

Our understanding of Roman urbanism relies on evidence from a few sites that have been the subject of large-scale clearance or major excavation campaigns, such as Pompeii and Ostia, which are unrepresentative of typical Roman cities. Non-invasive survey approaches on a multitude of abandoned ancient urban sites in Italy and elsewhere are now rapidly changing our approach to the Roman city. This short paper presents the outcome of the first high-resolution GPR surveys of complete Roman towns in Italy, Falerii Novi and Interamna Lirenas. We review the methods deployed and provide a brief overview of the results. We demonstrate how this type of survey has the potential to revolutionise archaeological studies of urban sites, while also challenging current methods of analysing and interpreting large-scale GPR data sets.

Recent work has demonstrated the value of GPR survey on Roman urban sites. Since 2015, we have deployed GPR on a large scale to generate high-resolution images of these two complete greenfield Roman towns in Lazio. Although such rapid data collection allows entire Roman cities to be mapped at an unprecedented level of detail, interpretation of these large data sets still relies largely on visual analysis and the manual digitisation of anomalies. These traditional, time-consuming interpretative methods are no longer able to exploit fully the potential of geophysical prospection, and here we propose possible ways forward. This includes most of all the integration of the GPR data with the data obtained from full scale geomagnetic prospection of both towns, and the integration with traditional and more innovative topographic and archaeological methods, such as surface survey, drone-based 3D modelling and aerial and satellite photography.

Finally, the presentation will also, in continuation of geophysical surveys carried out in recent years by this Belgo-British team, discuss the current and future strategies for stratigraphic contextualization, focused chronometry, 3D-visualisation of the physical landscape, creation of a DEM, erosion modelling and environmental reconstruction of both urban sites. A systematic augering program, targeted excavation and test-pitting are some of the operations under development.

Falerii Novi

Located approximately 50km to the north of Rome, along the ancient via Amerina, Falerii Novi has a walled area of 30.6 ha. The town was founded in 241 BC, following the destruction by Rome of the nearby Faliscan centre of Falerii Veteres. The exact status of the Republican town has remained uncertain. Occupation at Falerii Novi continued through Roman times and down to the early medieval period (sixth to seventh centuries AD). Today its remains lie mostly buried in open fields occupied only by a church and a former monastery, now used as a farm. Only the impressive ruins of the Republican town wall, in Medieval times used as boundary for the monastery domain, mark the emplacement of the former city. Falerii Novi has seen little excavation, with the exception of work in the early 19th century and a large trench excavated in the 1960s. As a consequence, our evidence for the town comes almost exclusively from non-invasive methods. It was one of the first Roman towns to be subjected to a complete fluxgate gradiometer survey coordinated by the BSR (Keay *et al.* 2000; Hay *et al.* 2010), providing a very clear plan of the entire intra-mural site and part of its northern suburb. It revealed the overall layout of the town and the original street grid and suggested how this original plan subsequently expanded up to the town walls (Millett 2007). The survey also revealed many details on housing and public buildings (e.g., the central forum complex), as well as a series of temples around the periphery of the town.

To complement the results of the magnetometry survey a new (almost) total coverage GPR survey (26.6 ha) was undertaken at Falerii in three summer seasons between 2015 and 2017, by a team from the universities of Ghent and Cambridge (Verdonck *et al.* 2020). The GPR network, towed by a quad, comprised 15 500MHz antennae, resulting in a vertical profile spacing of 0.125m. In order to meet sample density requirements, a second pass was made, reducing the transect spacing to 0.0625m, and achieving maximum resolution. After following a standard GPR data-processing workflow, including background removal, the output resulted in a series of time-slices, which map the GPR data as a series of images at successive depths below the surface. Detailed archaeological analysis

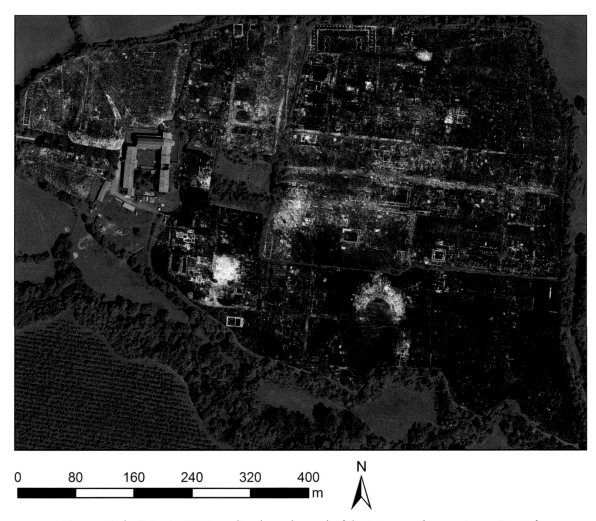

Figure 1. Falerii Novi: GPR time slice (sample area) of the intramural area, at an estimated depth of 0.75-0.80m.

and interpretation was possible only after first examining each separate time-slice individually, manually mapping anomalies in a GIS environment and interpreting the anomalies in terms of known or suspected Roman architectural features and forms. Work in progress, using both GIS and other techniques to explore true 3D visualisation for data analysis and interpretation of the GPR output, can still enhance the results from the field survey. Also, the important issue of the hugely time-consuming manual definition of the detected anomalies is currently being addressed by research into AI approaches exploiting the linear and orthogonal character of many anomalies, which usually represent walls, wall foundations and regular floor spaces in Roman towns.

The results from the Falerii Novi GPR survey show clearly that the high resolution of the data and the ability to distinguish features at different depths provide a much stronger foundation for understanding the town than was previously possible (Figure 1). GPR survey at Falerii Novi has revealed several previously unrecorded public buildings, such as a few temples and presumed sanctuaries (e.g., a huge porticus duplex), a *macellum* and a bath complex. Also, the contribution to the understanding of the city's dense domestic infrastructure is impressive, revealing some of the houses in great detail and suggesting a pattern for the early organisation of housing space in the city, connected with the original foundation plan. But, although the GPR data also help with a problem encountered with the magnetometer survey, providing a clearer view where in the magnetometer data building rubble masked structural detail, neither method is able to produce a complete picture of the buried archaeology. In some locations the fluxgate gradiometer survey produced a clearer image (e.g., the *tabernae* on the forum), underlining the by now well-known need to deploy complementary prospection methods and to fully integrate the results using image fusion and integrated interpretative mapping approaches.

In June 2021 a new phase of systematic prospection activity was initiated at the site, as part of a collaboration of the Ghent/Cambridge team with a team from the universities of Harvard and Toronto and the British School at Rome. Additional work within the intramural extent of the city has four principal objectives:

- to check the accuracy of the previous magnetometer and georadar surveys with respect to the position of the anomalies indicated and to obtain additional information about their stratigraphic contexts;
- to recover a sample of ceramic materials that should help refine the known chronology of occupation throughout the city, as well as the various types of activity that took place there;
- to gather data to clarify and reconstruct the physical landscape at the time of the city's foundation, as well as its subsequent transformation, by means of a digital elevation model (DEM) and erosion modelling;
- to recover samples of organic materials (e.g. pollen) from which information about the ancient environment at the site may be extracted.

To reach these objectives two minimally invasive methods were introduced:

1. a series of approximately 120 small (ca. 0.5 x 0.5 m) test pits dug to an estimated depth of 20-25 cm in a 50m grid systematically covering the ancient city allows to collect diagnostic artefacts and confirm correspondence between the upper archaeological layers and the results of the geophysical surveys undertaken in previous years.
2. a series of circa 50 manual augerings, systematically aligned along two sections (Figure 2) through the intra-mural city area allow to test the potential of the technique at Falerii Novi to yield minimally invasive stratigraphic, geological, and environmental information, with samples taken at different depths for radiocarbon dating, as well as other paleo-ecological and scientific analyses.

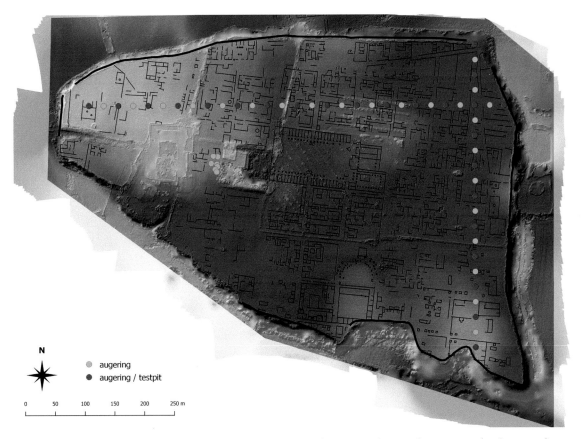

Figure 2. Falerii Novi: strategy of two sections with the location of manual augerings (and testpits) on a background of a LiDAR-based DEM and a map of archaeological structures derived from the magnetic survey.

Interamna Lirenas

The town of Interamna Lirenas was established as a Latin colony in 312 BC. Strategically placed along the via Latina, in a central position within the lower Liri Valley, it occupied a plateau controlling an important crossing over the river *Liris* (then navigable). Apart from some very scant remains of a bridge, a bath complex, an aqueduct and few cisterns, archaeological evidence had for long amounted to potsherds and tile fragments brought to the surface by ploughing (Hayes and Wightman 1984). This situation began to change in 2010, when a new fieldwork project was launched by Cambridge, in partnership with the BSR, the Italian Soprintendenza and the local Municipality (Bellini, Launaro and Millett 2014). A full-scale geomagnetic survey was carried out over the entire site (approx. 24 ha, 2010-12). Although this survey did not produce detailed plans of the buildings in the town, it provided key evidence for the layout of the street system and the location of the forum. In 2012-13, a GPR survey of a limited sector (50x50 m) near the forum led to the discovery of a hitherto unknown (roofed) theatre, whose entire plan has since been brought to light through excavation (2013-19).

Such discovery showcased the high potential of GPR and, in partnership with Ghent, the whole site was surveyed with the same GPR network employed at Falerii Novi (2015-17). This returned a remarkably crisp image of a very dense and articulated urban settlement, featuring a wide variety of buildings, in terms of both typology and size. Although this most likely reflected the situation in the 1st-2nd c. AD (see below), which resulted from considerable modification and rebuilding since the town's foundation, traces of the original (colonial) property plots can be discerned over the

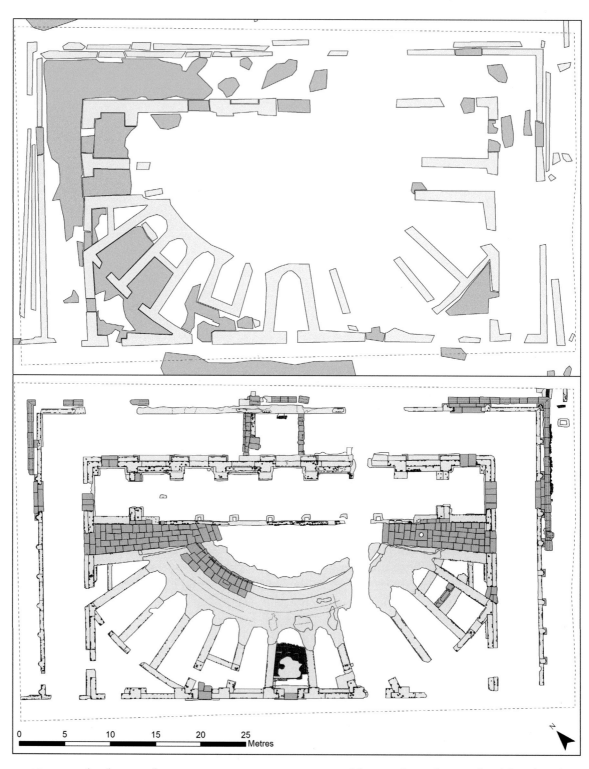

Figure 3. The theatre of Interamna Lirenas: interpretation of the geophysical anomalies (above) and excavation (below).

Figure 4. Intensity of occupation at Interamna Lirenas (AD 100-250) as indicated by the spread of commonware pottery.

entire plateau. Leaving aside public spaces, said plots appear to have belonged to two main types, defined by size, location and number (a situation resonant with Cosa, another Latin colony).

As in Falerii Novi, different techniques proved to be complementary: while streets feature way more prominently in the geomagnetic survey, building plans appear incomparably clearer and more detailed in the GPR. The efficacy of both techniques, however, may have petered out in some marginal areas, likely characterised by a deeper soil cover resulting from modern land transformations. Furthermore, although excavations have confirmed the remarkable level of accuracy of the geophysical survey, they also highlighted how its reach did not exceed c. 70 cm in depth (Figure 3).

A systematic survey of ploughsoil finds was also designed to map the chronological development of the town. Rather than attempting complete surface collection, our work combined two complementary approaches. First, all archaeological material from the ploughed surface was collected from a series of circular sample areas ($25m^2$ each) across the site and placed at c. 30m intervals to provide systematic coverage. Second, within each of them, a test pit was dug (c. 0.5m x 0.5m, 0.3m deep), the soil from which was sieved and all archaeological finds collected. Our aim was to ensure that we would recover reliable samples of material across the site. Earlier research had shown commonware pottery to be a better indicator of past occupation intensity in all periods, both at the site and in the surrounding territory (Launaro and Leone 2018). The spread of potsherds belonging to this class was thus plotted and processed to produce Kernel Density maps for each period. These maps not only highlighted that the area around the forum had been occupied without interruption from the late 4th c. BC until the 6th c. AD, but also showed how the peak of occupation at the town, reached in the late Republic, persisted well into the 3rd c. AD (Figure 4) – in contrast with ideas of a precocious decline (supposedly undergoing by the end of the 1st c. BC already) which had been put forward by earlier scholarship.

A new phase of research has since been launched in July 2020. Our overall approach is now being extended over parts of the extramural periphery of Interamna Lirenas and has already led to the discovery of what are likely to be structures pertinent to the town's river-port.

References

Bellini, G. R., Launaro, A. Millett, M. 2014. Roman colonial landscapes: Interamna Lirenas and its territory through Antiquity, in J. Pelgrom, T. D. Stek (eds) *Roman Republican Colonization. New Perspectives from Archaeology and Ancient History*: 255-275. Rome: Palombi.

Hay, S., Johnson, P. Keay, S., Millett, M. 2010. Falerii Novi: further survey of the northern extramural area. *Papers of the British School at Rome* 78: 1–38.

Hayes, J. W., Wightman, E. M. 1984. Interamna Lirenas: risultati di ricerche in superficie 1979-1981. In S. Quilici-Gigli (ed.) *Archeologia Laziale VI*: 137-148 Rome: Consiglio Nazionale delle Ricerche.

Keay, S. J., Millett, M., Poppy, S., Robinson, J., Taylor, J., Terrenato, N. 2000. Falerii Novi: a new survey of the walled area. *Papers of the British School at Rome* 68: 1–93.

Launaro, A., Leone, N. 2018. A view from the margin? Roman commonwares and patterns of distribution and consumption at Interamna Lirenas (Lazio). *Journal of Roman Archaeology* 31: 323-338.

Millett, M. 2007. Urban topography and social identity in the Tiber Valley, in R. Roth, J. Keller (eds) *Roman by Integration: dimensions of group identity in material culture and text* (Journal of Roman Archaeology Supplementary Series 69): 71–82. Portsmouth (RI): Journal of Roman Archaeology.

Verdonck, L., Launaro, A., Vermeulen, F., Millett, M. 2020. Ground-penetrating radar survey at Falerii Novi: a new approach to the study of Roman cities. *Antiquity* 94 (375): 705–723.

A multidisciplinary approach for characterizing the shallow subsoil of the Central Archaeological Area of Rome for geohazard assessment

M. Moscatelli[1], M. Mancini, F. Stigliano[1], M. Simionato[1], C. Di Salvo[1], G.P. Cavinato[1], S. Piro[2]

[1] CNR-IGAG, Istituto di Geologia Ambientale e Geoingegneria, Rome (Italy).
[2] CNR-ISPC, Istituto di Scienze del Patrimonio Culturale, Rome (Italy)
Massimiliano Moscatelli - massimiliano.moscatelli@igag.cnr.it

Introduction

In 2009, after declaring a state of emergency for the Central Archaeological Area of Rome following the adverse weather conditions of November and December 2008, the Government Commissioner and the Italian Department of Civil Protection (DPC) assigned the Institute of Environmental Geology and Geoengineering (IGAG) of the Italian National Research Council (CNR) to evaluate the geohazard level affecting the Central Archaeological Area of Rome (i.e., Palatine Hill, Roman Forum, and Colosseum). Research activities started up in February 2009 and were concluded in February 2011, with the valued contribution of the Archaeological Superintendence.

Methods

The first two phases of the project (between February and December 2009) were carried out mainly using information available from public Institutions and private companies (Figure 1). Three new thematic maps - on a 1:1,000 scale - were produced at the end of the second phase: (i) a geological and hydrogeological map with seven geological cross-sections, (ii) a susceptibility to instability map, and (iii) a seismic amplification susceptibility map. The third and final phase allowed to perform new field surveys (Figure 1) and involved several CNR research Institutes (apart from IGAG, scientific coordinator, among the others the Institute for Technologies Applied to Cultural Heritage ITABC, now Institute of Heritage Sciences ISPC, was involved) and University departments.

The new surveys (Figure 1) were planned to achieve two main goals: 1) the characterization of the archaeological layer, in order to (i) map the anthropic cover, and (ii) typify the archaeological layer in terms of physical and mechanical properties; 2) the characterization of the underlying geological bedrock, aimed to (i) map the geological units and typify the recognized lithotypes in terms of physical and mechanical properties, (ii) detect and monitor the water table position, and (iii) detect conditions potentially susceptible of instabilities and seismic amplification.

In the recent years, the research activities in the archaeological area continued with reference to the Colosseum. In 2016, a new campaign of geognostic investigations financed by the Special Superintendence involved the hypogea of the Colosseum, allowing for the acquisition of geological, geotechnical, and geophysical information of the recent alluvial fills of the Labicano ditch and the middle Pleistocene succession that house the foundations of the Colosseum.

The subsoil setting of the study area (less than 1 km^2) was preliminarily defined by means of already available 200 geotechnical boreholes. A new drilling campaign was also undertaken to investigate better the geological and archaeological layers of Palatine hill and surrounding areas. Twenty-five continuous coring vertical boreholes (total depth between 20 and 60 m) were drilled at the Palatine and five at the hypogea of the Colosseum.

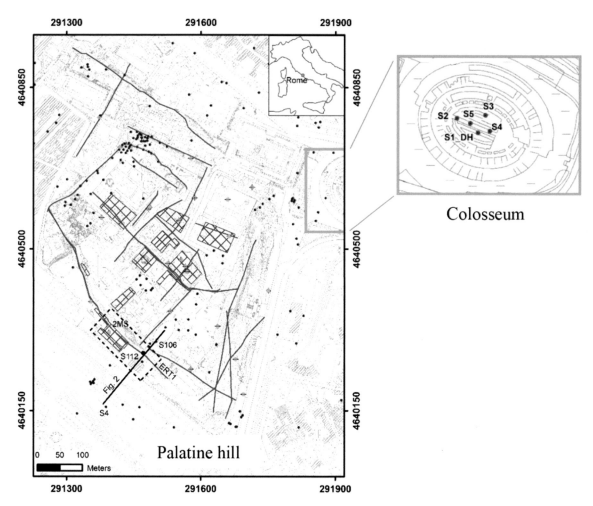

Figure 1. Location map of the available data. Legend - black points: previous boreholes; green points at Palatine hill and red points at Colosseum: new boreholes (2010 and 2016 surveys, respectively; S1-5: borehole; DH: down-hole); blue lines: ERT surveys; red checked areas: GPR surveys. The track of the geological cross-section and ERT1of Figure 2, and the position of the correlated boreholes are also reported. Dashed black box indicates the Schola Praeconum-Arcate Severiane zone.

All the boreholes crossed the anthropic layer, which ranges in thickness between 1 meter and more than 20 meters (at the Labicano valley). Several boreholes also crossed the network of tunnels dug in the tuff rocks underling the anthropic layer.

Lithological and stratigraphic logs of boreholes, integrated with information from local archaeological stratigraphy, allowed to strongly constrain i) the bottom surface of the archaeological layer, and ii) the boundaries between geological units.

Because almost no direct observation either of the geological bedrock or of the pre-anthropic/ anthropic contact is possible due to the thousand-year-old anthropic covering, an extensive geophysical survey was then planned to extend correlations all along the study area, and to characterize internal variability of subsoil units.

Twenty-four ERT (Electrical Resistivity Tomography) were performed at Palatine hill and Roman Forum. Resistivity field data were collected using different array configurations (Wenner-Schlumberger and Dipole-Dipole) and electrode spacing (from 1 to 10 m), obtaining different investigation depth (from about 8 to 80 m). In all cases, the resistivity values range from 10 to more than 1280 Ωm. As regards the archaeological layer, in general, relatively high resistivity values (>400 Ωm) are associated to voids and/or cemented conglomeratic walls, while low to moderate resistivity values (<400 Ωm) are related to anthropogenic silty-sandy backfill material.

With GPR (Ground Penetrating Radar) method a high-resolution data acquisition technique was adopted to reconstruct a global image of five areas. For the measurements, a 500 MHz bistatic antenna with constant offset, a 70 MHz monostatic antenna and a 35 MHz monostatic antenna, were employed. The horizontal spacing between parallel profiles at the site was 0.5 m, employing the 500 and 70 MHz antennas, and 1m employing the 35 MHz antenna. Some signal processing and representation techniques were used for data elaboration and interpretation. With the aim of obtaining a planimetric image of all possible anomalous bodies detected in the ground, the time-slice representation technique was applied using all field profiles. Amplitude of reflections recorded in the time-slices is mainly referable to the distribution of archaeological structures. High amplitude reflections can be referred to archaeological remains and, locally, to voids located in the anthropic layer. Low amplitude reflections can be related to the anthropic backfill and, in few cases, to the geological bedrock.

Integration of the geophysical prospections with geotechnical boreholes allowed to define a detailed geological model of the study area, in terms of buried topography, stratigraphy, geometries and lithotype distribution of natural geological and anthropogenic bodies.

By means of borehole correlations in the framework of the available geophysical surveys (mainly ERT), through lithofacies interpretation and within the current knowledge on the Rome Basin stratigraphy, several homogeneous geological domains were defined for which the internal complexity was resolved. This approach allowed the complex stratigraphic architecture of the near surface geology of the study area to be reconstructed (Mancini *et al.*, 2014). The geological bedrock of Palatine hill and surrounding areas is constituted by a Pliocene sandy-clayey unit of marine origin, the Monte Vaticano Formation (Funiciello and Giordano, 2008), that was drilled just south of the Palatine hill by the Circo Massimo exploration borehole in the 1930s (Signorini, 1939), for a total thickness of about 900 meters. The Quaternary complex covering the Monte Vaticano Formation is constituted by fluvial-palustrine and distal volcanic deposits belonging to several superimposed geological formations, which are mainly middle Pleistocene in age and show a multilayer vertical arrangement of paleo-valley fills. These fluvial-palustrine deposits form the backbone of Palatine and surrounding hills and were in turn carved by local tributaries of the Tiber River during the last Late Quaternary sea-level fall, giving rise to deep (up to 70-80 meters) and narrow alluvial valleys confined by steep slopes joining the hilltop plateaus: the Velabro, Labicano, and Murcia valleys to the west, east, and south of the Palatine hill, respectively. These valleys were filled mainly by clayey deposits in response to the Holocene sea-level rise and are presently buried by the anthropogenic deposits.

With regards to the anthropic layer, the buried morphology was reconstructed through the integration of ERT and GPR surveys, that were calibrated and constrained using geological cross-sections, borehole and archaeological stratigraphies (Figure 2). The basal surface of the anthropic layer was interpolated by means of geostatistical methods (Moscatelli *et al.*, 2014). In terms of composition, which is an important parameter when dealing with assessment of local seismic response, the anthropic layer was distinguished in: i) zones with dominant masonry remains, and ii) zones with dominant infill. Zones with dominant masonry consist predominantly of building remains, typically alternate with sandy-pebbly fill materials with a silty-clay or pozolan matrix. Zones with dominant infill are generally subordinate and consist of sandy-pebbles with a silty-clay or pozolan matrix. This classification was functional to the numerical modelling that was carried out to assess the local seismic effects of the archaeological area (see further details on this topic).

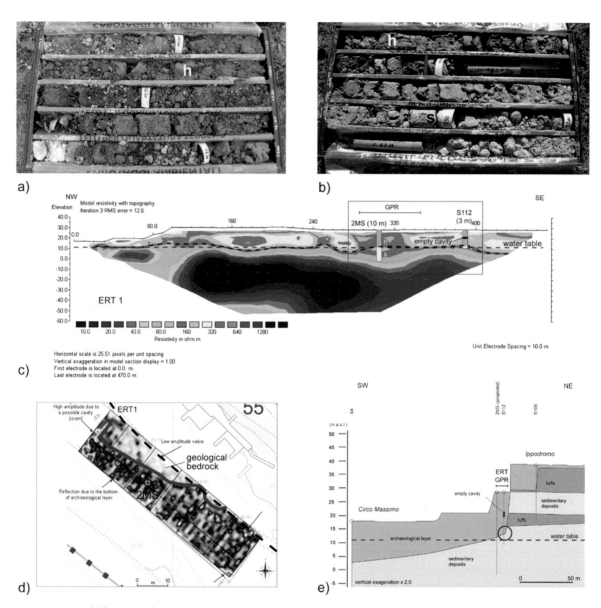

Figure 2. a) b) cores of the archaeological layer (h) from the borehole 2MS (Figure 1 for location; site elevation at 28.5m a.s.l.), between 10m and 15m (a) and 15m and 20m (b): weathered brick wall (10-14.1m) and concrete wall (14.1-16.4 m) remains are visible - the contact (green line) with the geological bedrock (s, sedimentary deposits) is visible in picture b) (red tubes are geotechnical samples); c) ERT 1 with interpretation of the probable contact between archaeological and geological layers (dashed black line) - red box indicate the Schola Praeconum-Arcate Severiane area, where the logs of geotechnical boreholes (in brackets the distance of projection in meters; h and s indicate the archaeological layer and the sedimentary deposits of the geological bedrock, respectively) and the extension of GPR survey are reported; d) example of a GPR amplitudes map, for 35 MHz antenna, at the averaged estimated depth of 15 m, showing the probable boundary (green line) between the archaeological layer and the geological bedrock (low amplitude zone) - track of ERT 1 and position of borehole 2MS are also reported; e) geological cross-section (for the track see Figure 1) showing the reconstructed contact between archaeological and geological layers, that is consistent with ERT and GPR data - the green circle indicates a toe of slope morphological setting, that is coherent with the GPR amplitude map imaged in d).

With specific reference to the Colosseum, geological and hydrogeological data allowed to reconstruct a preliminary 3D hydrostratigraphic model, with the final aim of establishing flood risk mitigation strategies for the defence of the archaeological heritage (Di Salvo *et al.*, 2020).

Without a doubt one of the most significant legacies of the Roman Empire, the Colosseum is considered one of the most famous monuments in the world. It receives the highest annual number of visitors to all cultural heritage sites in Italy (6.5 M in 2016: MiBACT, 2016). However, such a high flow of tourists has raised the exposure to hazards and, consequently, generated an increase in overall risk.

In fact, the area where the Colosseum was erected is in the valley floor of the ancient Labicano ditch. In this area the groundwater emerged, stagnating at the foot of the Velia hill, which made it necessary in historic times to reclaim the area through the construction of drains. In recent years, the local morphological conditions favourable to the accumulation of water persisted and the whole Central Archaeological Area has experienced numerous floods caused by intense storm events (Di Salvo *et al.*, 2017). During a storm on 20 October 2011, the hypogea of the Colosseum were quickly flooded by a copious amount of water that reached a depth of 6m from the lowest topographic level of the monument.

The model was helpful for detecting geological contacts, calculating volumes and thicknesses, and evaluating geometric relationships between hydrostratigraphic complexes. The 3D model, together with historical-archaeological information and observations, allowed for the development of a conceptual model describing the dynamics of groundwater and surface water inflows toward the Amphitheatre.

Moreover, geotechnical in situ and laboratory tests, active (Down-Hole, Cross-Hole, and MASW) and passive (noise measurements) geophysical surveys allowed us to define a subsoil model aimed at 1D and 2D numerical evaluation of the seismic response of the area (Pagliaroli *et al.*, 2014a).

The seismicity of the Roman area is considered of modest entity, compared to the whole context. However, over the more than 2500 years of its history, Rome has been affected by the earthquakes of the neighbouring seismogenic areas (essentially the central Apennines to the east and the volcanic complex of the Alban Hills to the southeast), recording a significant amount of damage to the historical and monumental heritage. Several authors have examined historical sources to build a site catalogue containing macroseismic resentments in the city of Rome during historical earthquakes (Galli and Molin, 2014, and reference therein). Among the others, according to historical accounts, the strongest seismic shaking ever felt in Rome occurred on September 9, 1349, when the sub-contemporary rupture of several Apennine seismogenetic structures caused one of the most devastating seismic sequences experienced throughout the recorded Italian history. Rome was violently struck, and this probably caused the collapse of the southern sector of the external ring of the Colosseum.

Investigation of the physical phenomena responsible for site effects at Palatine hill and surrounding areas shows that ground motion distribution is mainly controlled by 1D resonance phenomena and 2D effects associated with i) soft alluvial valleys bordering the hill, ii) superimposed paleo-valley fills, and iii) present-day topography and buried morphology of the anthropic layer. The performed numerical modelling (Pagliaroli *et al.*, 2014b) shows maximum amplifications of ground shaking, related to local geological and morphological settings, that can be significantly relevant for the monumental and archaeological heritages of this area, as many are highly vulnerable due to their great age.

Conclusions

In this work an integration of non-intrusive and intrusive methodologies is presented, aimed at the characterization of the subsoil of an archaeological area of extremely high value, such as the Central Archaeological Area of Rome. The application clearly demonstrates the potential for full integration of geological, geophysical, and archaeological methodologies to better characterize the geological and anthropic layers in archaeological areas. As a matter of fact, results show that: 1) both buried topography and internal complexity of the archaeological layer are detectable; 2) main boundary surfaces between geological units are laterally traceable; 3) hydrogeological information can be interpreted in the light of archaeological information and field observations to assess the susceptibility to flooding of monuments; 4) geological, geophysical and geotechnical information can be integrated and modelled to assess the susceptibility of the different sectors of the archaeological area to amplify the seismic shaking in the event of an earthquake.

These findings are the fruit of repeated prospecting campaigns, which have been for the most part non-intrusive. The cost of these campaigns is undoubtedly high, and this case study might seem difficult to replicate in other archaeological areas. When considering, however, the value that these monuments have for the capacity of worldwide tourist attraction (without considering their inestimable value from a historical and cultural point of view), the cost of these campaigns becomes negligible when compared with the risk of their damage or loss.

References

Di Salvo, C., Ciotoli, G., Pennica, F., Cavinato, G. P. 2017. Pluvial flood hazard in the city of Rome (Italy), *Journal of Maps* 13: 545–553.

Di Salvo, C., Mancini, M., Cavinato, G. P., Moscatelli, M., Simionato M., Stigliano F., Rea, R., Rodi, A. 2020. A 3D Geological Model as a Base for the Development of a Conceptual Groundwater Scheme in the Area of the Colosseum (Rome, Italy). *Geosciences* 10 (7): 266. doi.org/10.3390/geosciences10070266

Funiciello, R., Giordano, G. 2008. Note illustrative della Carta Geologica d'Italia alla scala 1:50.000, Foglio 347 Roma. APAT-Servizio Geologico d'Italia, Rome. Available at: https://www.isprambiente.gov.it/Media/carg/note_illustrative/374_Roma.pdf. (Accessed: 10 June 2021)

Galli, P., Molin, D. 2013. Beyond the damage threshold: the historic earthquakes of Rome. *Bulletin of Earthquake Engineering* 12 (3): 1277-1306. doi.org/10.1007/s10518-012-9409-0

Mancini, M., Marini, M., Moscatelli, M., Pagliaroli, A., Stigliano, F., Di Salvo, C., Simionato, M., Cavinato, G. P., Corazza A. 2014. A physical stratigraphy model for seismic microzonation of the Central Archaeological Area of Rome (Italy). *Bulletin of Earthquake Engineering* 12: 1339–1363. doi.org/10.1007/s10518-014-9584-2

Ministero dei Beni e delle Attività Culturali e del Turismo MiBACT. 2016. *Ministero dei Beni e delle Attività Culturali e del Turismo. Direzione Generale Bilancio. Ufficio statistica. 2016-Visitatori e Introiti di Musei, Monumenti e Aree Archeologiche Statali, Tavola 8. Gli Istituti Museali Più Visitati*. Available at: http://www.statistica.beniculturali.it/Visitatori_e_introiti_musei_16.htm (Accessed: 8 February 2018).

Moscatelli, M., Piscitelli, S., Piro, S., Stigliano, F., Giocoli, A., Zamuner, D., Marconi, F. 2014. Integrated geological and geophysical investigations to characterize the anthropic layer of the Palatine hill and Roman Forum (Rome, Italy). *Bulletin of Earthquake Engineering* 12: 1319–1338. doi.org/10.1007/s10518-013-9460-5

Pagliaroli, A., Quadrio, B., Lanzo, G., Sanò, T. 2014a. Numerical modelling of site effects in the Palatine Hill, Roman Forum and Coliseum Archaeological Area. *Bulletin of Earthquake Engineering* 12: 1383–1403. doi:10.1007/s10518-013-9436-5

Pagliaroli, A., Moscatelli, M., Raspa, G., Naso, G. 2014b. Seismic microzonation of the central archaeological area of Rome: results and uncertainties. *Bulletin of Earthquake Engineering* 12: 1405–1428. doi:10.1007/s10518-013-9480-1

Signorini, R. 1939. Risultati geologici della perforazione eseguita dall'AGIP alla mostra autarchica del minerale del Circo Massimo di Roma. *Bollettino della Società Geologica Italiana* 58: 60–63.

Acquisition, integration and interpretation of multiple GPR data sets in urban areas, as part of the ERC Rome Transformed project

S. Piro, D. Zamuner,
T. Leti Messina, D. Verrecchia

Istituto di Scienze del Patrimonio Culturale, ISPC CNR, Rome.
Salvatore Piro - Salvatore.piro@cnr.it

Introduction

Important research and technical issues are related to prospection in urban areas to locate subsurface cavities and/or archaeological remains and the production of hazard mapping. In many cases, cavities, such as subsidence features, voids and collapses represent disruptions to the geometry of an originally near-horizontal layered system. Geophysical techniques can be employed to identify feature geometries by observing contrasts in physical properties, but utilities, structures and surficial debris can all interfere with instrument measurements.

The critical phase of the geophysical survey in urban areas is the interpretation of the collected data and the characterization of the degree of confidence in the interpretations. Urban subsoil often consists of many layers which attest the history of a place, preserving in essence records of alternating phases of construction and destruction. The shallow subsurface of modern cities contains reams of pipes, cellars, wells, cavities, tunnels, graves and foundation walls of former houses, churches and town fortifications. Underneath the tarmac of city roads and the paving stones of town squares layers of sand and gravel are criss-crossed with modern fibre optic and telephone cables and century old sewer pipes mixed with the debris of brick buildings.

The most promising non-destructive geophysical prospection method for use in urban areas is Ground Penetrating-Radar (GPR). GPR measurements are less affected by the presence of metallic structures than magnetometry and they result in the largest amount of data of all commonly employed near-surface geophysical methods, providing detailed three-dimensional information about the subsurface (Trinks *et al.* 2009; Piro and Goodman 2008).

While geophysical prospection is generally considered as the attempt to locate structures of archaeological interest, in many cases, when applied in urban centres, this attempt could fail due to the effect and disturbances caused by recent man-made structures in the subsoil, which veil any signal related to structures of archaeological interest. Modern underground structures (such as bars and slabs of reinforced concrete, metallic pipes, cables and associated trenches and building debris) occurring at a shallow depth often display a stronger contrast in physical properties relative to the surrounding subsoil than less well-expressed archaeological structures, which often are buried at greater depth (Piro and Goodman 2008; Piro and Zamuner 2016).

Challenges for GPR prospection in city centres lie in the large number of obstacles present in the urban environments. Traffic islands, metallic drain covers, lamp posts, buildings, trees and parked vehicles cause irregular survey geometries, holes in the surveyed area and disturbing anomalies in the GPR measurements.

Archaeological prospection in urban centres

As indicated above, archaeological geophysical prospection in urban areas can be limited or affected by recent man-made structures in the sub-soil.

At the complex site considered in the Rome Transformed Project, a series of GPR surveys employing different frequencies were carried out. For the field measurements two different GPR SIR3000 and SIR4000 Systems (GSSI) were used, the first equipped with a 400 MHz antenna with constant offset and the 70 MHz monostatic antenna, the second equipped with digital dual frequency antenna with 300/800 MHz. The 400 MHz antenna was a compromise between depth penetration to about 2-3 m and resolution of features in the order of 0.15 – 0.20 m in order to define the archaeological features of interest. The 70 MHz antenna was employed to investigate at a depth penetration more than 2-3 m and with a resolution more than 0.30 m. The 300/800 MHz antenna was employed to investigate simultaneously two different depth ranges, 0-3 and 0-5 m, with different resolution.

Acquisition was made using a high-resolution approach in which parallel profiles were recorded very closely across the site. Signal processing, image processing, and visualization techniques have been used in conjunction with data modelling, elaboration, and interpretation of the recorded subsurface amplitudes (Goodman and Piro 2013). With the aim of obtaining a planimetric vision of all possible anomalous bodies the time-slice representation technique was applied using all processed profiles (Goodman and Piro 2013; Piro *et al.* 2020). Time-slices are calculated by creating 2-D horizontal contour maps of the averaged absolute value of the wave amplitude from a specified time value across parallel profiles.

The basic radargram signal processing steps included: (i) post processing pulse regaining; (ii) DC drift removal; (iii) data resampling; (iv) band-pass filtering; (v) background filter and (vi) migration.

All the GPR profiles were processed with GPR-SLICE v7.0 Ground Penetrating Radar Imaging Software (Goodman 2020).

In the present paper the surveys made with GPR to investigate different sites in the area of S. Giovanni in Laterano, Scala Santa and Santa Croce in Gerusalemme, as part of the ERC-funded Rome Transformed project (2019-2024) are presented and discussed. The acquisition discussed took place in March, June, July, October 2020 and January 2021. The geophysical team of ISPC CNR had previously investigated the selected area indicated in Figure 1.

The Rome Transformed Project aims to enhance the knowledge of Rome place in cultural change across the Mediterranean world by mapping political, military and religious changes to the Eastern Caelian from the first to eighth centuries AD (Haynes *et al.* 2020). An important aspect of the project strategy is the employment of different geophysical methods (GPR and ERT) suitable for deep stratigraphic urban investigations.

San Giovanni in Laterano

A small part of the area around S. Giovanni in Laterano Basilica was previously investigated (2012) by team members employing GPR equipped with 400 and 70 MHz antenna. Many traces of possible archaeological structures were found, in particular recorded in the north-western area of the Piazza S. Giovanni in Laterano (Piro *et al.* 2020).

The aim of the new GPR surveys is on the one hand to verify what has already been identified with the previous GPR surveys to locate Roman and high-medieval age remains and on the other to extend the area to be investigated towards the south of the square. The survey at San Giovanni in Laterano square was undertaken between in early March 2020 and a comprised a team from CNR ISPC.

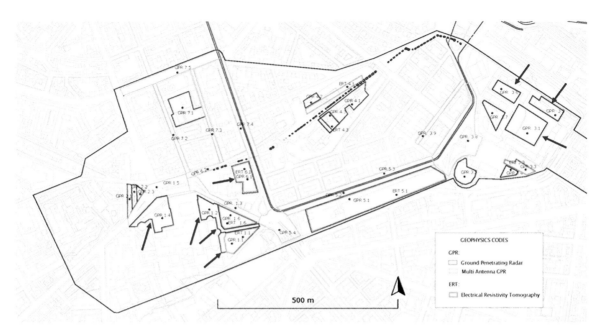

Fig. 1 - GPR surveys Area GPR 1.1, 1.2, 1.4, 1.6, 3.1, 3.2, 3.6 and 6.1

Figure 1. Location of the area investigated with the GPR systems (blue arrows).

For the measurements a GPR SIR3000 (GSSI), equipped with a 400 *MHz* (GSSI) bistatic antenna with constant offset, a 70 *MHz* (Subecho Radar) monostatic antenna and a SIR4000 (GSSI) system equipped with dual frequency antenna with 300/800 MHz were employed.

The horizontal spacing between parallel profiles at the site was 0.50 *m*, employing the four antennas. Radar reflections along the transects were recorded continuously, with different length, and horizontal stacking was set to 3 scans.

In the area outside and around the Basilica (Piazza S. Giovanni in Laterano e Piazza Giovanni Paolo II) a total of 876 adjacent profiles across the site were collected alternatively in forward and reverse directions employing the GSSI cart systems equipped with an odometer. All radar reflections within the 90*ns* for 400 *MHz* antenna, 125-140 and 230 *ns* for 70 *MHz* antenna (two-way-travel) time windows and a depth range of 2-4 m with 300/800 MHz antenna were recorded digitally in the field as 16 and 32 bit data and 512 samples per radar scan.

A nominal microwave velocity of about 0.12 *m/ns* was determined from fitting hyperbolas to the raw field data. This was used in estimating a penetration depth from the GPR survey.

With the aim of obtaining a planimetric vision of all possible anomalous bodies, the time-slice representation technique was applied using all processed profiles showing anomalous sources up to a depth of about 2.5 m, (Haynes *et al.* 2020; Goodman and Piro 2013).

Reflection amplitude maps (time slice data sets) were generated by spatially averaging the squared wave amplitudes of radar reflections in the horizontal as well as the vertical. The squared amplitudes were averaged horizontally every 0.25 *m* along the reflection profiles 3 *ns* (for 400 *MHz* antenna) and 6 *ns* (for 70 *MHz* antenna) time windows (with a 10% overlapping of each slice). The resampled amplitudes were gridded using the inverse distance algorithm with a search radius of 0.75 *m*.

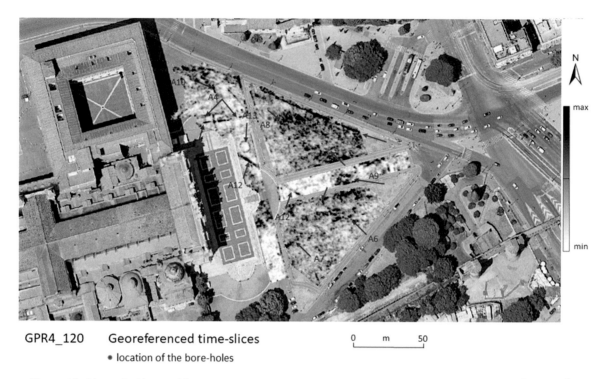

GPR4_120   Georeferenced time-slices

• location of the bore-holes

0   m   50

Figure 2A. Piazza S. Giovanni in Laterano. GPR area 1.1, 1.2, 1.4. 400 MHz antenna; GPR time-slices at the estimated depth of 1.1 – 1.3 m.

Mt. Mario Italy, EPSG 3004   0   m   50

Figure 2B. Piazza S. Giovanni in Laterano. GPR area 1.1, 1.2, 1.4. 70 MHz antenna; GPR time-slices at the estimated depth 1.7 – 2.0 m.

Figure 2A shows the time slices corresponding to the estimated depth of 1.1 – 1.3 m obtained with the 400 MHz antenna in the San Giovanni in Laterano square; Figure 2B shows the time slices corresponding to the estimated depth of 1.7 – 2.0 m obtained with the 70 MHz antenna. The area located in the S. Giovanni Paolo II square has been investigated also with the 300/800 MHz antenna, and the obtained results are equivalent with those obtained employing 400 MHz. The GPR images are characterized by the presence of structures with different dimension and linear and/ or circular shapes.

In Figure 2A, the size of the anomalies identified with 400 MHz antenna are approximate: (A1), (A2) and (A3) are not visible; (A4) traces of linear anomalies (possible remains of walls) with dimension 31.3 x 1.2 m. (A6) portion of a rectangular anomaly; (A7) portion of a linear anomaly with dimension 5.5 x 1.8 m; (A9) anomaly with dimension 5.3 x 6.0 m; (A10) anomaly due to probable utility; (A11) confirmed anomaly with low intensity; (A12) utility (related to ACEA service) with dimension 60.3 x 1.5 m; (A13) circular anomaly with diameter 7.30 and size 1.1 m.

Scala Santa garden

The survey in the garden of the Scala Santa was undertaken in July 2020 and a comprised a team from CNR ISPC.

Scala Santa Aree GPR 6.1 – July 2020
Results of GPR 300 MHz surveys.

0   m   30

Figure 3. Scala Santa. GPR area 6.1. 300 MHz antenna. GPR time-slice at the estimated depth 1.7 – 1.9 m.

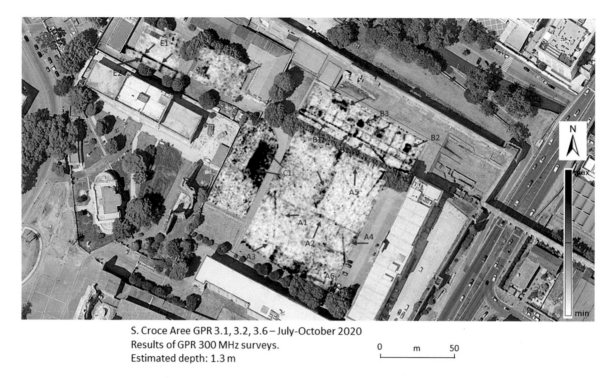

S. Croce Aree GPR 3.1, 3.2, 3.6 – July-October 2020
Results of GPR 300 MHz surveys.
Estimated depth: 1.3 m

0   m   50

Figure 4. S. Croce in Gerusalemme. GPR area 3.1, 3.2, 3.6. 300 MHz antenna. GPR time-slices at the estimated depth of 1.3 m.

For the measurements a GSSI SIR4000 system equipped with dual digital frequency antenna with 300/800 MHz was employed. The horizontal spacing between parallel profiles at the site was 0.50 m, employing this dual frequency antenna. Radar reflections along the transects were recorded continuously, with different length and horizontal stacking set to three scans.

In this area a total of 108 adjacent profiles across the site were collected alternatively in forward and reverse directions employing the GSSI cart system equipped with an odometer. All radar reflections within a depth range of 2-4 m with 300/800 MHz antenna were recorded digitally in the field as 32-bit data and 512 samples per radar scan. Time slice data sets were generated by spatially averaging the squared wave amplitudes of radar reflections in the horizontal as well as the vertical.

Figure 3 shows the time slices corresponding to the estimated depth of 1.7 – 1.9 m obtained with the 300 MHz digital antenna. The GPR images are characterized by the presence of few anomalies with different dimension and linear shapes. One of these anomalies, indicated by the blue dotted line, seems to have the same direction of those structures located through archaeological excavations made along the adjacent road.

Santa Croce in Gerusalemme

The surveys at Santa Croce in Gerusalemme were undertaken in early July 2020, early October 2020 and mid-January 2021 by a team from CNR ISPC.

For the measurements a GSSI SIR3000 equipped with a 70 MHz monostatic antenna and a SIR4000 system equipped with digital dual frequency antenna with 300/800 MHz were employed. The horizontal spacing between parallel profiles at the site was 0.50 m, employing this dual frequency antenna. Radar reflections along the transects were recorded continuously, with different length and horizontal stacking set to 3 scans.

In this area a total of 455 adjacent profiles across the site were collected alternatively in forward and reverse directions employing the GSSI cart system equipped with an odometer. All radar reflections within the 140 ns (twt) time window for 70 MHz antenna and a depth range of 2-4 m with 300/800 MHz antenna were recorded digitally in the field as 16- and 32-bit data and 512 samples per radar scan.

Time slice data sets were generated by spatially averaging the squared wave amplitudes of radar reflections in the horizontal as well as the vertical.

Figure 4 shows the time slices corresponding to the estimated depth of 1.3 *m* obtained with the 300 MHz digital antenna. The GPR images are characterized by the presence of many anomalies with different dimension and linear shapes with a regular geometrical location. Area A: A1 circular anomalies with size 1.3 m and diameter 6.0 m; A2, linear anomaly with dimension 1.2 m x 41.0 m. A3, anomaly with dimension 1.4 x 48.5 m; A4, anomaly with dimension 3.1 x 23.4 m; A5, anomaly with dimension 1.4 x 5.0 m (3 segments of perpendicular utilities ?). Area B: this area is characterized by the presence of many linear and perpendicular anomalies with the following dimensions: B1, 1.2 x 78 m, and 1.2 x 19.8 m (the shorter); B2, 3.7 x 6.4 m and B3, 3.2 x 2.2 m. Area C: at this depth we observe a big high reflected surface with dimension of 109 square meters. Area D-E: at this depth, in D area we do not observe any clear anomalies, but a diffused small reflection. In E area we observe two kinds of linear and perpendicular anomalies with dimensions, E1, 0.9 x 12.0 m and E2, 0.7 x 4.3 m.

GPR survey at the selected areas produced significant and fruitful results which demonstrate that when appropriately targeted and analysed GPR can be successfully undertaken for archaeological purposes in complex urban environments.

Acknowledgments

This project has received funding from the European Research Council (ERC) under the European Union's Horizon 2020 research and innovation programme (grant agreement no. 835271).

The project is based also on a wide consortium of partners in Rome who have both facilitated the research as well as share all information related to the study area: Soprintendenza Speciale Archeologia Belle Arti e Paesaggio di Roma; Comune di Roma – Sovrintendenza Capitolina ai Beni Culturali; Musei Vaticani.

References

Goodman, D., Piro, S. 2013. *GPR Remote sensing in Archaeology*. Berlin: Springer. DOI 10.1007/978-3-642-31857-3

Goodman, D. 2020. *Manual of GPR Slice v7.0*, Ground Penetrating Radar Imaging Software.

Haynes, I. P., Liverani, P., Kay, S., Piro, S., Ravasi, T., Carboni, F. 2020. Rome Transformed: Researching the Eastern Caelian C1-C8 CE (Rome). *Papers of the British School at Rome* 88: 354-357.

Piro, S., Goodman, D. 2008. *Integrated GPR data processing for archaeological surveys in urban area. The case of Forum (Rome, Italy)*. 12th International Conference on Ground Penetrating Radar, June 16-19, 2008, Birmingham, UK. Proceedings Extended Abstract Volume.

Piro, S., Zamuner, D. 2016. Investigating the urban archaeological sites using Ground Penetrating Radar. The cases of Palatino Hill and St John Lateran Basilica (Roma, Italy). *Acta IMEKO* 5(2): 80-85. DOI: 10.21014/acta imeko/v5i2.234 .

Piro, S., Haynes, I. P., Liverani P., Zamuner, D. 2020. Ground Penetrating Radar Survey in the Saint John Lateran Basilica, in L. Bosman, I. P. Haynes, P. Liverani (eds) *The Basilica of Saint John Lateran to 1600*: 52-70. Cambridge: Cambridge University Press. DOI: 10.1017/9781108885096

Trinks, I., Karlsson, P., Biwall, A., Hinterlaitner, A., 2009. Mapping the urban subsoil using ground penetrating radar – challenges and potentials for archaeological prospection, *ArchaeoSciences revue d'archeometrié* 33 (suppl.): 237-240.

Integrated GPR and laser scanning of Piazza Sant'Anastasia, Rome

E. Pomar, S. Kay, P. Campbell, K. Vuković

The British School at Rome (Italy)
Elena Pomar - e.pomar@bsrome.it

Introduction

The application of non-invasive techniques for archaeological research, in particular geophysical prospection and 3D-modelling, have been at the forefront of the research conducted by the British School at Rome (BSR), in particular at the sites of Portus and San Giovanni in Laterano. The techniques have been used for the investigation of both large-scale sites as well as documenting complex buildings, in many cases as a precursor to archaeological excavations. However, when sites are embedded in a complex urban landscape, non-intrusive methods offer the only possibility for a detailed investigation of the subsurface. The survey at Piazza Sant' Anastasia falls within this context.

The church of Sant'Anastasia and the adjoining piazza occupy an area of great importance in the topography of ancient Rome. Seated between the Palatine and the Aventine hills, the area is adjacent to the Circus Maximus and very close to the Forum Boarium (for the ancient topographical context of the study area, see among others Coarelli 2008).

The aim of the survey was to locate archaeological features underneath the piazza with Ground-Penetrating Radar (GPR) and to record the current layout of the area, dominated by the church façade, with high-resolution 3D laser scanning. The ultimate objective was to combine the data in a shared 3D environment capable of representing the diachronic evolution of the site.

Historical and archaeological context

The basilica of Sant'Anastasia is one of the oldest churches in Rome, its construction dated to the time of Pope Damasus in the 4th century AD. The structure was rebuilt and renovated numerous times up until the major Baroque intervention of 1722 (Krautheimer 1937: 54-60). Prior the restoration of the façade in the 17th century the basilica was accessed by a staircase, which is visible in 16th century illustrations (Krautheimer 1937: 43).

The Paleochristian church was built over a group of Roman structures, dated from the 1st to the 4th century AD, which were used as foundations of the new building (Whitehead 1927: 405-410). Below the nave a series of Roman *tabernae*, rectangular in shape, opened onto a road which runs underneath the right-hand aisle, parallel to the modern via dei Cerchi. Traces of an upper floor, likely belonging to an *insula*, are still visible incorporated into the wall of the church towards the Palatine hill (Figure 1). Excavations beneath the church were conducted between 1857 and 1863 (Detlefsen 1859; Bergau 1863) and the structures were later studied by Whitehead (1927) and subsequently by Krautheimer (1937). More recently scholarship in this immediate area of the Valle Murcia has focused on the relation to the primordial legends of Rome, as some scholars place the Lupercal cave in the surrounding area of Sant'Anastasia (Carandini 2008; Vukovic 2018).

Figure 1. Lefthand side of the church of Sant'Anastasia, towards the Palatine hill (BSR Photographic Archive, John Henry Parker Collection, jhp-1922).

Challenges and strategy

The location and intense urbanisation of the surrounding area of Piazza Sant'Anastasia have meant that no previous excavations have taken place. Clues as to the continuation of structures beneath the piazza are provided by the excavations beneath the church, however, to test this hypothesis a geophysical survey of the piazza was undertaken.

Among the geophysical instruments commonly applied in archaeology, the context and the aim of the project suggested GPR as the most suitable technique. The urban nature of the site, paving, modern services and depth excluded the application of magnetometry. Electrical Resistivity Tomography (ERT), whilst suitable for urban environments, was excluded due to practical issues (primarily the paving of the piazza and access to the church from the piazza).

Together with the geophysical prospection, a 3D laser scan of the piazza and the church façade, was undertaken. The aim of the laser scanning was to collect a 3D point cloud, allowing in the first instance the creation of a Digital Terrain Model (DTM) to topographically correct the GPR data. The piazza lies at the foot of the Palatine Hill with a slope from east to west with a variance up to 3m, therefore topographical correction was required of the GPR data to allow a correct interpretation.

Finally, in order to house all the data within a shared 3D environment, including the geophysical and laser scan data and digitised plans of the structures underneath the church, a topographical survey was carried out with a Global Position System (GPS).

Survey method

The geophysical prospection was carried out using a GSSI SIR-3000 coupled with 200MHz and 400MHz antennas. The complementary use of two antennas at different frequencies was intended to maximise the spatial resolution at different penetration depths (Conyers 2013). The GPR survey was conducted in the accessible areas of the piazza (1108m²) along parallel NE-SW traverses at a regular distance of 0.5m for the 200MHz antenna and 0.25m for the 400MHz antenna.

The 3D laser scanning was undertaken using a Leica RTC 360 scanner with fixed targets, the position of which were subsequently recorded with a GPS (Leica GS18). The GPS was also used for establishing the geophysics grid.

The data was processed in dedicated software (ESRI ArcMap for the GPS data, GPR-Slice for the GPS data and Leica Cyclone for the laser scanner) and then subsequently combined for additional post- processing. The DTM generated from the laser scanner point cloud was exported as an ASCII file (.xyz) and added into the GPR software for the topographical correction of the GPR data. The processed GPR data were subsequently exported as an ASCII file (.xyzirgb) and combined with the georeferenced point cloud and the digitize plan from Whitehead (1927: Plate 11) in a shared 3D environment. A future step, in collaboration with the Parco Archeologico del Colosseo, will be to integrate the new 3D model and geophysics data with a 3D model of the excavated standing structures beneath the church.

Results

Beyond the shallowest level, where the modern services underneath the piazza were recorded, the GPR located a series of linear features of potential archaeological interest. A comparison of the GPR

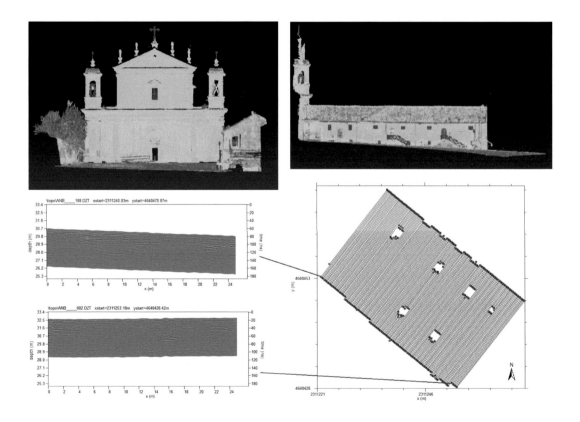

Figure 2. Composite image showing the slope of the piazza and the topographical correction applied to the GPR data.

time-slices with the cartographic sources (including the modern *Carta Tecnica Regionale*, the '*Forma Urbis*' of Lanciani and the plan of the Roman *tabernae*) in the Geographic Information System (GIS) environment allow for several archaeological interpretations to be made of the data.

At a depth of approximately 0.9m below the pavement of the piazza an elongated NE-SW feature, parallel to the church façade, was recorded up to a depth of -1.6m. The feature likely locates the remaining footings of the staircase which originally gave access to the church before its Baroque restoration.

The clearest results regarding the Roman structures were collected at a greater depth, between -1.6m and -3.5m with the 200MHz antenna. A series of high amplitude reflectors indicate the continuation of the *tabernae* westward, beyond the façade of the church (Figure 2). Two large parallel features are oriented NW-SE, in accordance with the Roman topographic organization of the area. Of these, the southern feature (26m in length) is aligned with the road onto which the *tabernae* opened, whilst the northern feature (35m in length) is aligned with the wall that divided the *tabernae* from the rear rooms. The two linear anomalies delimit a rectangular space whose width, 14.5m, corresponds with the length of the rooms in the plan of Whitehead. Some internal divisions of the space are also indicated by other reflectors.

An analysis of the geophysical data indicates that the structures beneath the piazza follow the same plan of those under the church. The area occupied by the rear rooms of the *tabernae* probably lie just beyond the survey grid to the north, where benches and potted plants prevented accessibility.

The integration of all the data types provides a virtual three-dimensional representation of the historical evolution of the site, including the Roman buildings, the church with traces of its initial building phase and the present appearance of the area (Figure 3).

Figure 3. Results of the 200MHz GPR prospection between 2.5m and 2.7m below ground level.

Figure 4. 3D representation of the full dataset collected by the project.

Conclusions

The centre of Rome is characterised by a deep stratigraphy of archaeological sequences, whose traces are often hidden by the modern urban landscape. The survey at Sant'Anastasia, through the virtual combination of the excavated structures with the standing building and the buried remains, offers an example of the advantages derived from the application of non-invasive techniques in an urban context. In such a complex situation, where invasive excavation is unfeasible, non-intrusive methodologies offer a clear and advantageous opportunity to generate new important data.

Acknowledgements

The research is led by the BSR in collaboration with the Parco Archeologico del Colosseo with thanks to the Director Dott.ssa Alfonsina Russo. The authors are grateful to Taneisha Tamayo for her assistance in the field and to Mr. Peter Smith, for both his kind support and generous funding of the research.

References

Bergau, A. 1863. Scavi sotto la chiesa di S. Anastasia. *Bullettino dell'Instituto di Corrispondenza Archeologica per l'anno 1863*: 113-116.
Carandini, A. 2008. *La casa di Augusto dai 'Lupercalia' al Natale*. Bari/Rome: Editori Laterza.
Coarelli, F. 2008 *Roma*. Bari/Rome: Laterza.
Conyers, L. B. 2013. *Ground-Penetrating Radar for Archaeology*. Plymouth: AltaMira Press.
Detlefsen. D. 1859. Scavi sotto la chiesa di S. Anastasia. *Bullettino dell'Instituto di Corrispondenza Archeologica per l'anno 1859*: 139-142.
Krautheimer, R. 1937. *Corpus Basilicarum Christianarum Romae. The Early Christian Basilicas of Rome (IV-IX cent). Vol. 1*. Città del Vaticano: Pontificio Istituto di Archeologia Cristiana.
Vuković, K. 2018. The topography of the Lupercalia. *Papers of the British School at Rome* 86: 37-60.
Whitehead, P.B. 1927. The church of S. Anastasia in Rome. *American Journal of Archaeology* 31.4: 405-420.

GPR survey in the Punic harbour of La Martela (El Puerto de Santa Maria, Spain) and the methodology used for the processing and archaeological visualisation of the data

J.A. Ruiz Gil, L. Lagóstena Barrios, J. Pérez Marrero, P. Trapero,
J. Catalán, I. Rondán-Sevilla, M. Ruiz Barroso.

Departamento de Historia, Geografía y Filosofía, Facultad de Filosofía y Letras, Unidad de Geodetección, Análisis y Georreferenciación del Patrimonio, Universidad de Cádiz (Spain)

José-Antonio Ruiz Gil - jantonio.ruiz@uca.es
Lázaro Lagóstena Barrios - lazaro.lagostena@uca.es
Jenny Pérez Marrero - jenny.perez@uca.es
Pedro Trapero Fernández - pedro.trapero@uca.es
Javier Catalán González - javier.catalan@uca.es
Isabel Rondán Sevilla - isabel.rondan@uca.es
Manuel Ruiz Barroso - manuel.ruiz@uca.es

Introduction

This contribution aims to advance a suitable methodology to develop rigorous historical-urbanistic cartography, with a high-value use for research using the results of multichannel GPR exploration over a large area (Goodman and Piro 2013).

Although the process of geophysical exploration with multichannel equipment uses interpolation as a mechanism to visualize the data, these results offer enough quality to achieve valid archaeological planimetry (Conyers 2008).

On the one hand, this cartography we generated, should be recognized by the academic community as having a similar value to traditional excavation. On the other hand, it should not be directly related to high-quality discovery, -Virtual Archaeology- but to non-invasive, or non-destructive research. The goal is to get as close as possible to the representation of archaeological reality, without visual modification motived by the desire to obtain a more attractive result, as usually happens in processing intended for Virtual Reality or increased Reality products.

For characterizing results, the representation obtained must meet the following parameters:

- Georeferencing and Geographic Information Systems compatibility.
- Data in 2D and in 3D.
- Measurable information.
- Indications of construction phases, functions, abandonment phases, and other related elements.

This will allow for comparative historical analysis with archaeological remains and urban structures known by traditional research methods, indications on construction phases, functionalities, and abandonment phases.

To illustrate the procedure, we use the case of the Punic port of La Martela, located in the municipality of El Puerto de Santa Maria, Cádiz, Spain (Figure 1). The site was located by ground responses visible in aerial photographs (Lagóstena and Ruiz 2021), whose archaeological remains have been studied exclusively by geophysical techniques. The site in question is directly related to the Phoenician site of Castillo de Doña Blanca, located a short distance to the north. This discovery

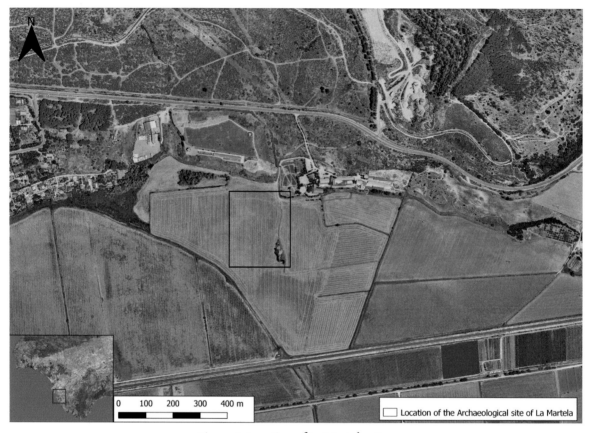

Figure 1. Location of La Martela site.

adds to the new approaches to the knowledge of ancient harbours and port structures, such as the recent research of Goiran *et al.* (2014) and Majchzack *et al.* (2017).

The main objective is to demonstrate that, thanks to geophysics, we can have scientific support that allows us to investigate several historical aspects of the settlement, contributing to the methodological development of non-invasive research, without having to apply traditional archaeological intervention. Also, as a secondary objective, we want to discuss the workflow of data collection, the correct processing of the data, as well as some post-processing of our own. All this is carried out to create an archaeological drawing from the geophysical information and the results of the interpretative data, to generate historical information.

Materials and methods

The multi-channel georadar equipment used in the present research (Stream-X IDS) is composed of a set of 15 antennas (16 channels) with a central frequency of 200 MHz separated by 12 cm each. The equipment scan band has a width of 2 m. The set has a lifting system that allows us to correctly adapt the height of the antenna to the prospected terrain, and a trailer system that allows a maximum prospecting speed of 15 kilometres per hour. The antennas are connected to a central control unit (DAD) where the data is collected. From the central unit, the system connects to a topographic precision odometer and the laptop with the software One Vision for data acquisition and GRED HD (2019) for post-processing the data. The georadar has been configured for this work with an exploration depth of 80 ns, 512 samplings per sweep (@ 512 Sample / Scan), an average propagation speed of 10 cm/ns and a GPS + PPS positioning system. The position is geo-located by a differential GPS Leica GNSS GS14 antenna with a CS15 controller, which exported the data from the receiver to the control unit in NMEA format every 0.2 sec (5 Hz) (Leica 2019).

Our method began with the survey conditions carried out on a sloped surface. We prospect in the line of depth and we found most of the remains in a flat area, concentrated in the centre of the exploration. In the nearest area to the Phoenician archaeological site of Castillo de Doña Blanca, the ground slopes drastically down and continues as a flat area, a paleo-channel representing the ancient path of Guadalete river, at two metres above sea level (Borja and Díaz del Olmo 1994; Ruiz and Pérez 1995, Caporizzo *et al.* 2021). The area where the main remains detected have been detected lies on a little promontory over the paleo-channel.

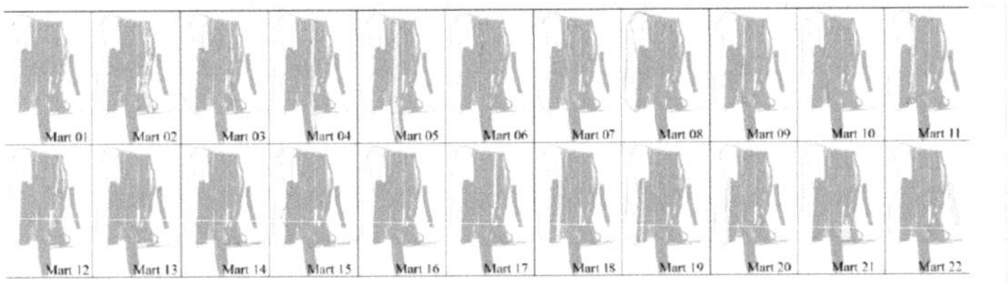

	Plot-Projects	Surface (^{2}m)	Perimeters	Time (min)	Averages (^{2}m/h)
1	2016.07.14_001	1.826,74	1.463	20	5480,205
2	2016.07.14_002	6.718,07	3.972	20	20154,216
3	2016.07.14_003	2.288,53	1.548	43	3193,30047
4	2016.07.26_001	4.973,97	1.941	43	6940,41907
5	2016.07.26_002	8.050,77	2.037	11	43913,2636
6	2016.07.26_003	1.228,42	1.157	10	7370,502
7	2016.07.26_004	1.894,44	1.012	13	8743,54615
8	2016.07.26_005	162,019	2.218	6	1620,19
9	2016.09.12_001	2.621,35	1.633	24	6553,365
10	2016.09.12_003	1.026,29	1.156	6	10262,87
11	2016.09.12_004	4.650,48	2.603	38	7342,86474
12	2016.09.12_005	2.575,01	2.011	28	5517,86786
13	2016.09.12_006	567,268	437	6	5672,68
14	2016.09.12_007	404,651	429	5	4855,812
15	2016.09.12_008	460,398	413,1	6	4603,98
16	2017.07.27_001	812,874	414	5	9754,488
17	2017.07.27_002	2.278,89	451	45	3038,516
18	2017.07.27_003	3.121,38	522	46	4071,36783
19	2017.07.28_001	2.182,59	514	28	4676,96786
20	2017.07.28_002	3.108,23	551	41	4548,62488
21	2017.07.28_003	1.845,72	376	50	2214,858
22	2017.07.28_004	2.585,85	594	37	4193,26541
Total		55.383,94	27.452	531	6258,07232

Figure 2. Information about GPR surveys.

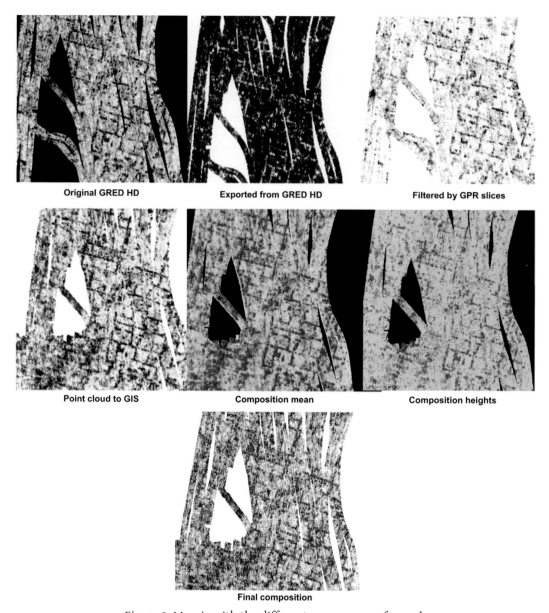

Figure 3. Mosaic with the different processes performed.

In general, GPR exploration was carried out following a north-south orientation, though in an alternating direction in each pass. These passages presented the width of Stream X, overlapping between them linearly in an irregular manner as a consequence of the length of each project. The position of the equipment moved with the vehicle that followed the track of the previous course of the cart and that was helped visually with a manual positioning by wooden stakes. Obstacles, such as trees or large stones, were avoided by a slight detour, which modified the line of the pass. In addition, the radar had to contend with the fact that the field was ploughed, which caused in some cases movements adjusted to the morphology of the soil (Figure 2).

Results

We have experimented with different graphic resources to obtain, with maximum rigour, digital cartography that effectively represents the evidence of archaeological elements below the surface. As shown in Figure 3, the different processes explored, already explained in detail in the previous section, provide images with slight variations. It is precisely these changes that define

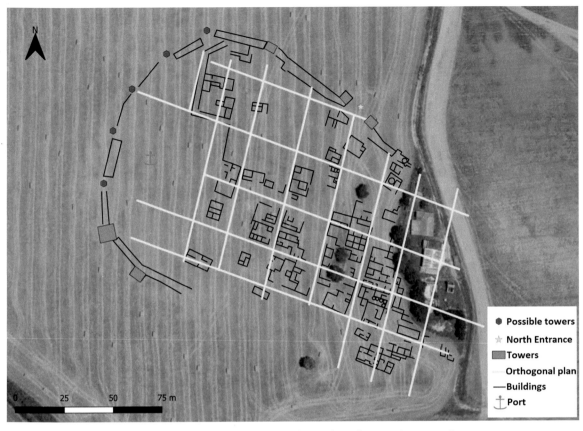

Figure 4. Archaeological plan of La Martela site based in Lagóstena and Ruiz 2021.

an image for archaeological use. The images are obtained by applying filters to the raw data, first in GRED HD images, which will later be exported (Figure 3.2) in a format compatible to receive the GPR SLICES filters (Figure 3.3). This result will be processed in GIS proprietary formats in the form of a point cloud (Figure 3.4). Finally, we have shown two possibilities to improve the image resolution (Figures 3.5 and 3.6).

Once we have a graphic representation useful for historical purposes, we move on to analyse the images (floor plans) and radargrams (longitudinal and transversal profiles) in combination. This method allows us to make decisions about the archaeological nature of the alterations detected. For this process, we use Geographic Information Systems (GIS), which helps us with the process of elaborating the historical cartography based on the analysis of the results obtained with the GPR. The visual proposal of the resulting exploration, once subjected to this methodology, is shown in the following figure (Figure 4). The final result is a rigorous planimetry of the archaeological remains, with academic value, which allows the historical analysis of the urban elements through traditional procedures (spatial organisation, dimension of the elements, comparative analysis and so on) (Lagóstena and Ruiz 2021).

Conclusions

To conclude, our main objective is historical and heritage knowledge using scientific techniques and methods such as GIS or geophysics. For us, these geophysical studies and the whole process involved could be of similar value to traditional historical and archaeological tools. This is why our team is made up of historians and archaeologists as well as engineers and computer scientists. Within the non-invasive research approach, this case study has enabled us to obtain a vector picture with all the potential of historical-archaeological modelling (delimitation of wall units, characterisation of the walls, quantitative and comparative analysis of the building units, etc.).

The enormous relevance of the contribution made in this paper is evident thanks to the method and techniques used. A Punic city port whose spatial and urban integrity covers an area of three hectares has been uncovered and is possibly a good case study for further research from the perspective of non-invasive techniques. All this is in the context of scarce investment of economic resources and in a very limited timeframe. The efficiency and effectiveness of non-invasive research is a fact in the field of history and archaeology, not to mention heritage management.

La Martela provides evidence for human activity concerning riverine evolution from the 5th to 4th centuries BC. Archaeologists propose the existence of a port on the SE side of Doña Blanca. This Punic port could be *cothon* type, and most likely continued in use to the end of the city, which was itself probably related to the blocking of the port as sediments accumulated to an unsustainable level. It is a new pre-Roman settlement discovery to be set alongside Gadir/Cadiz, Castillo de Doña Blanca and Cerro del Castillo.

La Martela settlement was built on a riverbank in front of the Doña Blanca site, inside the bay of Cadiz. The mouth of the Guadalete river was a great lagoon with marshes closed to the sea by a sand barrier. The fill of fluvial sediments was made in the same way by the two main rivers of the Cadiz Gulf: Guadalquivir and Guadalete. In both cases, the fluvial deposits were aligned in SE to NW orientation and continued to the coast (Caporizzo *et al.* 2021).

Before this research, only six hand-drilled boreholes had been by G. Hoffmann (1988). These probes were not geo-referenced so they must be located approximately to the SE of Doña Blanca towards the Guadalete paleochannel. This led to the identification of three 'anthropic formations interspersed in the clay-sandy stratigraphic series' (Borja and Díaz del Olmo 1994: 198).

In summary, the non-invasive survey has produced results of historical value. In this sense, we observe that:

- Archaeological remains are not visible at 'La Martela'.
- There is no written or cartographic information to indicate the location of the site.
- Only the non-invasive survey has provided proof of the port's existence.
- The evidence offered by our research assumes the existence of a new archaeological record.
- This record, plan and section in 3D, can be interpreted also with traditional plans and sections.

References

Borja Barrera, F. and Díaz del Olmo, F. 1994. Paleogeografía post-flandriense del litoral de Cádiz. Transformación Protohistórica del paisaje de Doña Blanca, in A. Morales, E. Roselló (eds). *Castillo de Doña Blanca, Archaeo-environmental investigations in the Bay of Cádiz, Spain (750-500 B.C.)* British Archaeological Reports International Series, 193: 185-198. Oxford: Archaeopress.

Caporizzo, F., Gracia, J., Aucelli, P. P. C., Barbero, L., Martín-Puertas, C., Lagóstena, L., Ruiz, J. A., Alonso, C., Mattei, G., Galán-Ruffoni, I., López-Ramírez, J. A., Higueras-Milena, A. 2021. Late-Holocene evolution of the Northern Bay of Cadiz from geomorphological, stratigraphic and archaeological data. *Quaternary International* 602: 92-109. https://doi.org/10.1016/j.quaint.2021.03.028

Conyers, L. B. 2008. Ground-penetrating radar for landscape archaeology: Method and applications, in S. Campana, S. Piro (eds) *Seeing the Unseen*: 245-255. London: CRC Press

Goiran, J. P., Salomon, F., Mazzini, I., Bravard, J. P., Pleuger, E., Vittori, C., Boetto, G., Christiansen, J., Arnaud, P., Pellegrino, A., Pepe, C., Sadori, L. 2014. Geoarcheology confirms location of the ancient harbour basin of Ostia (Italy). *Journal of Archaeological Science* 41: 389-398.

Goodman, D., Piro, S. 2013. *GPR remote sensing in archaeology.* London: Springer.

GRED HD, 2019, viewed 30 May 2019. https://idsgeoradar.com/products/ground-penetrating-radar/stream-x.

Hoffmann, G. 1988. *Holozänstratigraphie und Küstenlinienverlagerung an der andalusischen Mittelmeerküste*. Berichte aus dem Fachbereich Geowissenschaften der Universität Bremen 2. Bremen: Universität Bremen.

Lagóstena Barrios, L., Ruiz Gil, J. A. 2021. El puerto romano de Gades: nuevos descubrimientos y noticias sobre sus antecedentes, in L. Chioffi, M. Kajava, S. Örmä (eds) *Il Mediterraneo e la Storia III. Documentando città portuali*: 249-264. Rome: Quasar.

Leica. 2019. viewed 30 May 2019. https://leica-geosystems.com/es-es/products/total-stations/controllers/leica-viva-cs15-and-cs10

Majchczack, B. S., Schneider, S., Wilken, D., Wunderlich, T. 2017. Multi-method prospection of an assumed early medieval harbour site and settlement in Goting, island of Fohr (Germany), in B. Jennings, C. Gaffney, T. Sparrow, S. Gaffney (eds) *AP2017: 12th International Conference of Archaeological Prospection*: 249-251. Oxford: Archaeopress.

Ruiz Mata, D. Pérez, C. 1995. *El poblado fenicio del Castillo de Doña Blanca (El Puerto de Santa María, Cádiz)*, El Puerto de Santa María: Ayuntamiento del Puerto de Santa María.

4D with accuracy: why bother?

A. Schmidt[1,2,3], T. Sparrow[2], C. Gaffney[2], V. Gaffney[2], A. S. Wilson[2],
R. A. E. Coningham[1]

[1] Durham University, Department of Archaeology (UK)
[2] University of Bradford, School of Archaeological and Forensic Sciences (UK)
[3] Dr Schmidt - GeodataWIZ (D)
Armin Schmidt - A.Schmidt@GeodataWIZ.com

Abstract

Old photographs can be used with photogrammetric methods to reconstruct 3D models of past sites and monuments. Adding more recent data from laser-scanning allows building a 4D time-lapse record that can be enhanced further by incorporating 3D geophysical subsurface data, for example from Ground Penetrating Radar (GPR) surveys. Such comprehensive data sets facilitate accurate analysis that can highlight changes, either due to damage from natural disasters or as stylistic developments. The data collected form the Durbar Square in Bhaktapur, Nepal, highlight the benefits of this approach.

Keywords

TIME-LAPSE, 3D, RECONSTRUCTION , GPR, LASER-SCANNING

Introduction

For the communication of archaeological and other Cultural Heritage results to the public, 3D representations have long been very effective tools. In recent years, advances in information technology even facilitated 3D time-lapse animations based on different phases of sites and monuments (e.g. for the ship burial site in Gjellstad, Norway; https://www.gjellestadstory.no/). Although most such representations are based on archaeological results, they do not require high fidelity as large parts of such reconstructions are usually hypothetical. Would a full and accurate 4D recording (i.e. 3D with an additional time dimension) of archaeological data provide additional benefits, given that recording and data management would require more effort and possibly even new work paradigms?

All archaeological data are inherently 4D, as they capture objects, features and landscapes and try to put these into a chronological sequence with the aim of revealing their temporal evolution. Even a dated single object (e.g. a coin) is recorded in 4D as it was not in its find locations in the distant past, is presumed to have stayed there after its deposition and was excavated more recently. Various methods have been devised to optimise excavations and their recording to obtain as much information as possible. These include site maps, section drawings, context identifications and tools such as the Harris matrix (Harris 1979) to link such data together. During the post-excavation analysis these data are then conceptually assembled into a 4D narrative. Alternatively, several groups have demonstrated how 3D recording of surfaces during excavation, capturing all excavation phases separately and digitally, can be used as the basis for a complete 4D data set (Aspoeck and Fera 2015; Larsson *et al.* 2015; Schneidhofer *et al.* 2017), from which other, conventional products (e.g., section plans) could be derived, if desired.

Case study

We present here the benefits of compiling a 4D data set from a heritage site. This is based on the combination of work from two projects: 'Curious Travellers: Visualising Heritage' (http://www.visualisingheritage.org/CT.php) (Wilson *et al.* 2019) and 'Reducing Disaster Risk by Evaluating the

Figure 1. The Vatsala Durga temple (Bhaktapur Durbar Square, Nepal) after its destruction in the 2015 Gorkha earthquakes.

seismic Safety of Kathmandu's Historic Urban Infrastructure' (Coningham *et al.* 2019; Davis *et al.* 2019). One of the heritage sites that was investigated in these projects is the historic centre of the city of Bhaktapur in the Kathmandu valley, Nepal. Bhaktapur was the seat of a Newar kingdom in the 15th century and the city maintained an assemblage of many medieval buildings. It forms part of the Kathmandu Valley World Heritage Site and is one of the country's major tourist attractions. However. in 2015, the very strong Gorkha earthquakes (8.1 Ms) killed approximate 9000 people in Nepal and destroyed large parts of the country, damaging 1600 national heritage monuments (e.g., temples, palaces), of which 257 collapsed completely, including a large number in Bhaktapur (Figure 1). Due to the great importance of many of these temples for daily ritual life and their immense value for the tourism industry, the Department of Archaeology, Government of Nepal, decided to reconstruct approximately 200 of these destroyed monuments, starting as soon as possible after the earthquakes.

Although Nepal had made considerable efforts documenting its Cultural Heritage assets, not all information was created with the intention of delivering reconstruction accuracy. In most cases architectural examples were captured in illustrative drawings (e.g. Korn 2014) or as photographs with artistic quality (e.g. Gutschow and Basukala 2011). The current projects therefore aimed to provide additional and accurate details. The Gorkha earthquakes created two methodological epochs: for the pre-earthquake period photogrammetric reconstruction (Structure from Motion) was based on old photographs, while existing and reconstructed buildings of the post-earthquake period were captured with a handheld laser scanner. These two data sets formed the basis of a subsequent 4D evaluation. They were further contextualised with geophysical Ground-Penetrating Radar (GPR) data from a detailed investigation of the area's subsurface.

Figure 2. Bhaktapur Durbar Square, Nepal. Point cloud reconstructed from many images (blue triangles).
The Vatsala temple is seen on the right.

Results

The Curious Travellers project developed a framework for collecting old photographs of different heritage sites with the aim to reconstruct digitally some of the monuments using photogrammetric processing. This was achieved by (a) creating a data-mining infrastructure that was able to scrape the Internet (web and social media) for nearly one million images (Coningham *et al.* 2019; Wilson *et al.* 2019) and (b) establishing a crowd sourcing infrastructure to collect contributions from members of the public as part of a citizens-science project. This approach inevitably favoured those sites and monuments that appealed to tourists, for example Palmyra, in Syria, St Benedict, in Italy and Bhaktapur, in Nepal (Figure 2) (Wilson *et al.* 2019).

Sites and monuments that were depicted in a sufficiently large number of photographs were selected for processing and Structure-from-Motion algorithms were used to derive 3D digital models. Due to the considerable photographic coverage of the Vatsala temple in Bhaktapur its digital reconstruction was accomplished in detail (Figure 3) and even many other monuments in this historic square could be represented well (Figure 4). The data also demonstrate the limitations of this approach as tourist photographs only covered parts of the existing structures so that blank areas remained.

For the assessment of damages and changes after the earthquakes the site was subsequently recorded with a handheld laser scanner that uses an Inertial Measurement Unit (IMU) and Simultaneous Localisation and Mapping (SLAM) processing to obtain a spatially accurate point cloud of the whole square and its buildings (Figure 5). Additional images were captured separately to allow rendering of the resulting surface mesh. The scanned data also helped to position more accurately some of the photogrammetrically reconstructed pre-earthquake models.

After the Gorkha earthquakes geophysical surveys were undertaken in Bhaktapur's Durbar Square using Ground Penetrating Radar (GPR) with a single channel 500 MHz antenna using a

Figure 3. The Vatsala Durga temple (Bhaktapur Durbar Square, Nepal) as reconstructed form old photographs.

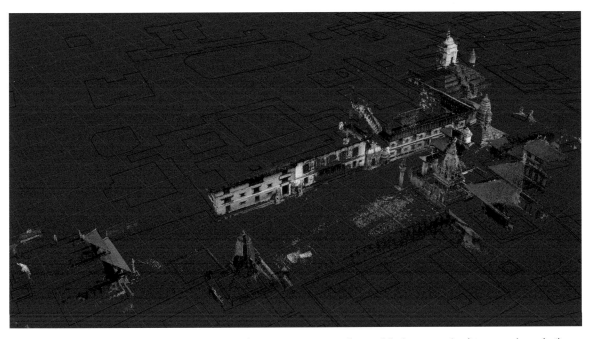

Figure 4. Bhaktapur Durbar Square, Nepal. Reconstructions form old photographs ('pre-earthquake').

spatial resolution of 0.05m × 0.35m (in-line sampling and line spacing, respectively) (Figure 6a). Additional targeted excavations in the Durbar Square showed a very complex stratigraphy, mainly formed by brick structures (e.g. wall remains and foundations) within a matrix of brick rubble. These conditions resulted in poor contrast for the GPR signal reflections and led to considerable noise in the collected data. Modern utilities (i.e., cables, water pipes and drainage) are crossing the square in many different directions at shallow depth (Figure 6b). At greater depth rectangular and rectilinear anomalies can be seen clearly (Figure 6a and b). They are interpreted as the foundations

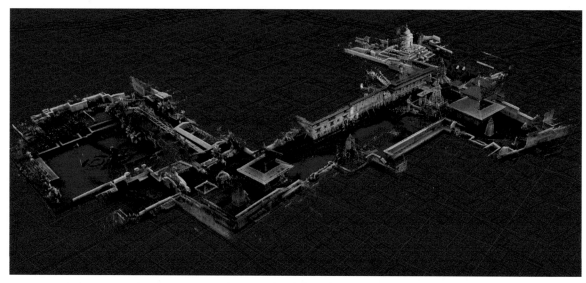

Figure 5. Bhaktapur Durbar Square, Nepal. Point cloud from laser-scanning ('post-earthquake'). The Vatsala temple is already partially rebuilt and the scaffolding is still visible.

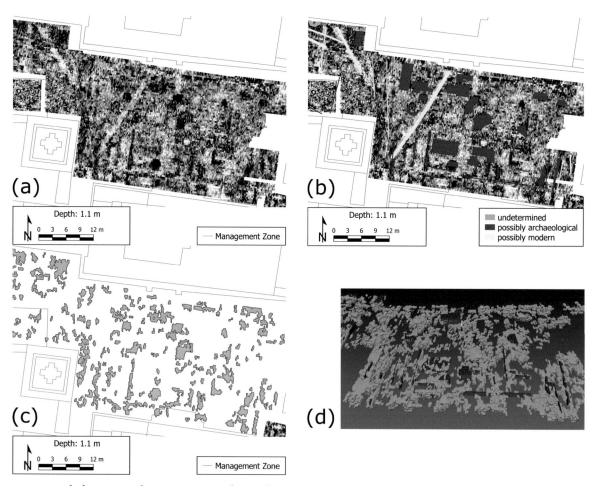

Figure 6. Bhaktapur Durbar Square, Nepal. GPR data at a depth of 1.1m. (a) greyscale representation (strong reflections in black); (b) interpretation diagram; (c) isolated anomalies after vector processing to reduce noise; and (d) 3D visualisation of processed GPR data (grey, see (c)) together with 3D interpretation of structures (red, see (b)).

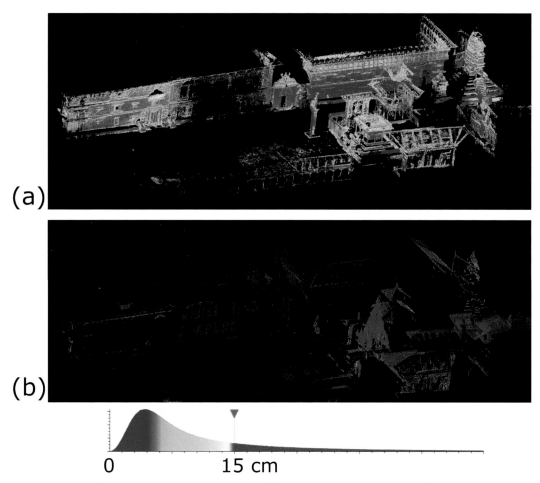

Figure 7. Bhaktapur Durbar Square, Nepal. Difference between the pre- and post-earthquake data. (a) Difference up to 0.15m; and (b) difference larger than 0.15m.

of a building located on the southern part of the square, as depicted in a watercolour by Henry Ambrose Oldfield in 1858 and destroyed by the strong earthquake in 1934. For the integration of GPR data with the 3D models of upstanding buildings a volumetric representation was required. On other sites where GPR noise from ground scattering is low and the contrast between archaeological features and the enclosing soil matrix is strong such representations can be achieved using isosurfaces (Schmidt *et al.* 2015; Verdonck *et al.* 2020) or point cloud rendering (Filzwieser *et al.* 2017). However, for the noisy data from Bhaktapur an alternative approach was required. Depth slices of 0.1m thickness were converted to polygons and processed with advanced spatial methods using PostGIS (Schmidt and Tsetskhladze 2013) (Figure 6c). The reduction in noise (Schmidt *et al.* 2020) allowed a visualisation based on the extrusions of processed polygons that compared well with the 3D visualisation of features from a manual interpretation (Figure 6d).

Evaluation

For a quantitative evaluation of the temporal changes between the 3D data sets the differences between pre- and post-earthquake data were calculated and are shown in Figure 7, separately for differences smaller (Figure 7a) and larger (Figure 7b) than 0.15 m. The section of the palace just west (left in Figure 7) of the central entrance has a far larger deviation than other parts, as it has moved outwards after the earthquakes and had to be supported with wooden beams. As the Vatsala temple was being reconstructed during the laser scanning the then still missing upper part of the temple dominates the difference-diagram for large deviations. There are several roof areas that

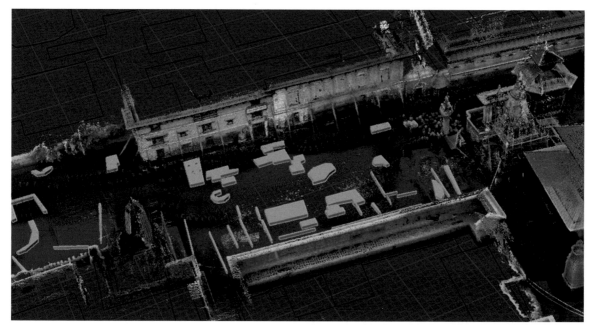

Figure 8. Bhaktapur Durbar Square, Nepal. Data from laser-scanning together with interpretation of GPR data (in grey).

also show large deviations, since roofs were difficult for the laser scan to reach from the ground. The data highlight the areas that moved and that are hence at greatest risk of further deterioration.

Combining above and below ground data (Figure 8) in a 3D environment (using the BlenderGIS Addon) allowed a far more realistic interpretation of the GPR anomalies since their spatial relationship to existing buildings became very clear.

Conclusions

For those time-lapse visualisations that are mostly used for presentations to the public it may be sufficient to base them on ground plans to estimate 3D representations of different phases. However, to be a detailed source of reference for conservators, architects and engineers, historic 3D data reconstructed from old photographs are far more useful. These allow accurate measurements of individual monuments and sites and of the relationships between upstanding and subsurface remains, recorded with geophysical methods. Furthermore, 4D data can be used for the documentation of changes between different phases and are hence important to assess damages as well as stylistic developments. The efforts of combining heterogeneous data into one accurate 4D representation are hence offset by the detailed analysis that becomes possible.

References

Aspoeck, E., Fera, M. 2015. 3D-GIS für die taphonomische Auswertung eines wiedergeöffneten Körpergrabes. *AGIT - Journal für Angewandte Geoinformatik* 1: 2-8.
Coningham, R. A. E., Acharya, K. P., Barclay, C. P., Barclay, R., Davis, C. E., Graham, C., Hughes, P. N., Joshi, A., Kelly, L., Khanal, S., Kilic, A., Kinnaird, T., Kunwar, R. B., Kumar, A., Maskey, P. N., Lafortune-Bernard, A., Lewer, N., McCaughie, D., Mirnig, N., Roberts, A., Sarhosis, V., Schmidt, A., Simpson, I. A., Sparrow, T., Toll, D. G., Tully, B., Weise, K., Wilkinson S., Wilson, A. 2019. Reducing disaster risk to life and livelihoods by evaluating the seismic safety of Kathmandu's historic urban infrastructure: enabling an interdisciplinary pilot. *Journal of the British Academy* 7(s2): 45-82.
Davis, C., Coningham, R., Acharya, K. P., Kunwar, R. B., Forlin, P., Weise, K., Maskey, P. N., Joshi, A., Simpson, I., Toll, D., Wilkinson, S., Hughes, P., Sarhosis, V., Kumar, A., Schmidt, A. 2019.

Identifying archaeological evidence of past earthquakes in a contemporary disaster scenario: case studies of damage, resilience and risk reduction from the 2015 Gorkha Earthquake and past seismic events within the Kathmandu Valley UNESCO World Heritage Property (Nepal). *Journal of Seismology* 24: 729-751.

Filzwieser, R., Olesen, L. H., Neubauer, W., Trinks, I., Mauritsen, E. S., Schneidhofer, P., Nau, E. Gabler, M. 2017. Large-scale geophysical archaeological prospection pilot study at Viking Age and medieval sites in west Jutland, Denmark. *Archaeological Prospection* 24(4): 373-393.

Gutschow, N., Basukala, B. 2011. *Architecture of the Newars: a history of building typologies and details in Nepal*. Chicago: Serindia Publications.

Harris, E. C. 1979. *Principles of Archaeological Stratigraphy*. London and New York: Academic Press.

Korn, W. 2014. *The traditional Newar architecture of the Kathmandu valley : the śikharas, a presentation of the different śikhara temple types found in the Kathmandu valley* (1st edn) Ratna art & architecture series. Kathmandu, Nepal: Ratna Pustak Bhandar.

Larsson, L., Trinks, I., Söderberg, B., Gabler, M., Dell'unto, N., Neubauer, W., Ahlström, T. 2015. Interdisciplinary archaeological prospection, excavation and 3D documentation exemplified through the investigation of a burial at the Iron Age settlement site of Uppåkra in Sweden. *Archaeological Prospection* 22(3): 143-156.

Schmidt, A., Dabas, M., Sarris, A. 2020. Dreaming of Perfect Data: Characterizing Noise in Archaeo-Geophysical Measurements. *Geosciences* 10(10): 382.

Schmidt, A., Linford, P., Linford, N., David, A., Gaffney, C., Sarris, A., Fassbinder, J. 2015. *Guidelines for the use of Geophysics in Archaeology: Questions to Ask and Points to Consider*. Namur: Europae Archaeologia Consilium (EAC).

Schmidt, A., Tsetskhladze, G. 2013. Raster was Yesterday: Using Vector Engines to Process Geophysical Data. *Archaeological Prospection* 20(1): 59-65.

Schneidhofer, P., Nau, E., Leigh McGraw, J., Tonning, C., Draganits, E., Gustavsen, L., Trinks, I., Filzwieser, R., Aldrian, L., Gansum, T., Bill, J., Neubauer, W., Paasche, K. 2017. Geoarchaeological evaluation of ground penetrating radar and magnetometry surveys at the Iron Age burial mound Rom in Norway. *Archaeological Prospection* 24(4): 425-443.

Verdonck, L., Launaro, A., Vermeulen, F., Millett, M., 2020. Ground-penetrating radar survey at Falerii Novi: a new approach to the study of Roman cities. *Antiquity* 94(375): 705-723.

Wilson, A. S., Gaffney, V., Gaffney, C., Ch'ng, Bates, R., Sears, G., Sparrow, T., Murgatroyd, A., Faber, E., Coningham, R. A. E. 2019. Curious Travellers: Repurposing imagery to manage and interpret threatened monuments, sites and landscapes. In M. Dawson, E. James, M. Nevell (eds) *Heritage Under Pressure – Threats and Solutions: Studies of Agency and Soft Power in the Historic Environment*: 107-122. Oxford, UK: Oxbow Books.

SITAR project. New approaches and methods for an open data archaeology of Rome

M. Serlorenzi, A. Cecchetti, A. D'Andrea, F. Lamonaca,
G. Leoni, R. Montalbano, S. Picciola

Soprintendenza Speciale per i Beni Archeologici di Roma (IT)
Mirella Serlorenzi - mirella.serlorenzi@beniculturali.it

SITAR (*Sistema Informativo Territoriale Archeologico di Roma - Archaeological Territorial Information System of Rome*) was launched in 2008 by the Soprintendenza Speciale per i Beni Archeologici di Roma in order to digitize and gather all the scientific data coming from the archaeological excavations and the geological research surveys carried out within the territory of Rome and Fiumicino. Its goal is to ensure the visibility, transparency and dissemination of the scientific data on archaeological excavations in the city of Rome: a digital registry dedicated to Rome's heritage, free for all to access and consult.

In 2020 a new website has been launched, in both Italian and English version, in order to provide more contents to the users (links, useful tutorials, resources, publications, etc) and to share approaches and methods with the community as a whole.

Since the beginning, SITAR main goal was to provide the community involved in the study and preservation of the archaeological and historical heritage of Rome with a useful support for processing archaeological data towards a shared urban co-planning approach. For this reason, unlike other similar experiences at national and international level, feature representation is no longer symbolic, but archaeological data are processed after an accurate georeferencing process carried out by professional archaeologists. To date, the system brings together different types of data, ranging from archival documentation to single archaeological remains found during rescue excavations.

After thirteen years from the development of the first web application, from May 2018 a system re-engineering was started, aimed at merging the three original applications into a single system. The infrastructure is now hosted on the GARR cloud and has a modular architecture, so that each service is allocated on specific virtual machines. This choice stems from a twofold requirement: on one hand, to optimize the response to individual requests, and on the other to ensure specific maintenance of the single services.

Among the main innovations the creation of a Digital Library is to be noted. This powerful tool allows users to explore SITAR documentation (maps, drawings, scientific reports), filtering the results through specific parameters. The new Digital Library is served by the ELK Stack: it uses the Elastic Search as search engine, Logstash for the index creation and Kibana to generate effective view on the data. Digital documents are indexed through an automatic OCR process and the system can retrieve the keywords used to search within every single document.

The final objective of this new re-factoring activity was to align SITAR with the FAIR DATA philosophy and therefore to guarantee an easy and well documented data acquisition. For this reason, SITAR data can now be acquired by any user through direct downloads thanks to the main available OPEN FORMATS (GEOJSON, GML2, GML3, KML; GEOTIFF, GEOTIFF8, SVG, CSV), through specific requests to the dedicated GEOSERVER instance or, at an upper level, thanks to REST API

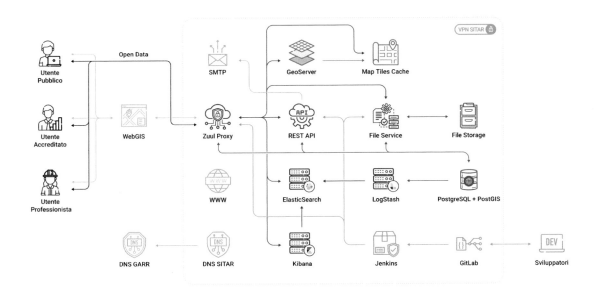

Figure 1.

services. The publication of the API allows the users to dynamically exploit the SITAR dataset, negotiating the protocol and the format according to its specific needs.

It is worth noting that SITAR database – that currently stores more than 6,000 excavations, 25,000 archaeological features and 100,000 attachments – adopts CIDOC CRM Archaeo as semantic model and the data have been extracted and represented in RDF, using the XML language. Moreover, as highlighted above, a new website was launched in 2020 in order to provide more content to the users (links, useful tutorials, resources, publications, etc.) and to share approaches and methods with the community as a whole.

References

Serlorenzi, M. 2011. SITAR. Sistema Informativo Archeologico di Roma, in M. Serlorenzi (ed.) *SITAR. Sistema Informativo Territoriale Archeologico di Roma. Atti del I Convegno (Roma, Palazzo Massimo, 2010)*: 9-29. Rome: Iuno Edizioni

Serlorenzi, M. (ed.) 2013. *ARCHEOFOSS Free, Libre and Open Source Software e Open Format nei processi di ricerca archeologica. Atti del VII Workshop (Roma, 11-13 giugno 2012)*. Archeologia e Calcolatori Supplemento 4. Rome: All'Insegna del Giglio

Serlorenzi, M. 2018. Accessibilità e diffusione del dato archeologico: l'esperienza del SITAR, in M. Arizza, V. Boi, A. Caravale, A. Palombini, A. Piergrossi (eds) *I dati archeologici. Accessibilità, proprietà, disseminazione (Roma - CNR, 23 maggio 2017)*. Archeologia e Calcolator 29: 31-40. Rome: All'Insegna del Giglio.

Serlorenzi, M., Jovine. I. (eds) 2013. *SITAR. Sistema Informativo Archeologico di Roma. Potenziale archeologico, pianificazione territoriale e rappresentazione pubblica dei dati. Atti del II convegno (Roma, 9 novembre 2011)*. Rome: Iuno Edizioni

Serlorenzi, M., Jovine, I. (eds) 2017. *SITAR. Sistema Informativo Archeologico di Roma. Pensare in rete, pensare la rete per la ricerca, la tutela e la valorizzazione del patrimonio archeologico. Atti del IV convegno (Roma, 14 ottobre 2015)*. Archeologia e Calcolatori, Supplemento 9. Rome: All'Insegna del Giglio.

Serlorenzi, M., Jovine, I., Boi, V., Stacca, M. 2016. Open Data in archeologia: una questione giuridica o culturale? in P. Basso, A. Caravale, P. Grossi (eds) *ARCHEOFOSS. Free, Libre and Open Source*

Software e Open Format nei processi di ricerca archeologica. Atti del IX Workshop (Verona 19-20 giugno 2014). Archeologia e Calcolatori Supplemento 8: 51-58. Rome: All'Insegna del Giglio.

Serlorenzi, M., Leoni, G. (eds) 2015. *SITAR. Sistema Informativo Archeologico di Roma. Il SITAR nella Rete della Ricerca italiana. Verso la conoscenza archeologica condivisa. Atti del III convegno (Roma, 23-24 maggio 2013)*, Archeologia e Calcolatori, Supplemento 7. Rome: All'Insegna del Giglio.

Serlorenzi, M., Lamonaca, F., Picciola, S., Cordone, C. 2012. Il Sistema Informativo Territoriale Archeologico di Roma: SITAR. *Archeologia e Calcolatori* 23: 31-50.

Serlorenzi, M., Lamonaca, F., Picciola, S. 2018. The SITAR Project: Web Platform for Archaeological Knowledge Sharing, in V. Apaydin (ed.) *Shared Knowledge, Shared Power Engaging Local and Indigenous Heritage*: 125-127. Cham: Springer.

Serlorenzi, M., Leoni, G., Lamonaca, F., Picciola, S. 2020. Il SITAR e le comunità degli utenti: un'infrastruttura culturale al servizio del patrimonio archeologico di Roma, in F. R. Cerami, M. L. Scaduto, A. De Tommasi (eds) *I bacini culturali e la progettazione sociale orientata all'Heritage Making. Tra politiche giovanili, innovazione sociale, diversità culturale*: 201-212. Palermo: All'Insegna del Giglio.

Marvellous metadata: managing metadata for the Rome Transformed project

A. Turner

Newcastle University (UK)
Alex Turner - alex.turner@newcastle.ac.uk

This paper explores the use of an integrated approach to the use of an Access database to manage the complex and varied metadata produced by the Rome Transformed project. The data can be divided into six broad categories, structural analysis, geophysical survey, environmental analysis, archival research, 3D modelling and visualisation and finally references. Each of the major categories includes a number of sub-categories to further enhance the management of metadata as can be seen below in Figure 1.

In addition to the metadata for the sub-categories, the standardised terms, instrument lists and contributing organisations and individuals are also stored within the database. Each of these is accessed from a central 'dashboard' to make it as easy as possible to locate and record the correct data category. The database is designed to be as user friendly as possible to encourage users to properly record their metadata. The failure of many projects to properly record the full range of metadata available, until the very end, can be overcome by adopting a simple and visual approach that utilises as many standardised lists as possible. Additional programmed functionality to reduce the amount of time needed to create a record greatly enhances the chances of a full set of metadata being recorded by the end of the project. It also enables the rapid extraction of data in the correct format for the creation of the final sustainable archive. The access to each sub-category is controlled by a basic map, Figure 2, that only includes the areas that are directly relevant to the sub-discipline. Simplifying access to the data in this way also provides a means of rapidly discovering areas where data is not fully recorded. The perceived tedious nature of the metadata recording process is sometimes reinforced by the impression that the data being recorded has no clear and obvious benefit to the overarching aims of the project.

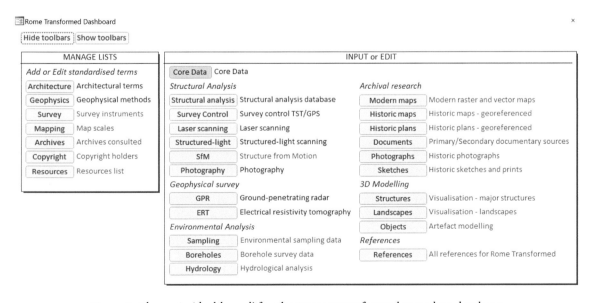

Figure 1. The main 'dashboard' for the Rome Transformed metadata database.

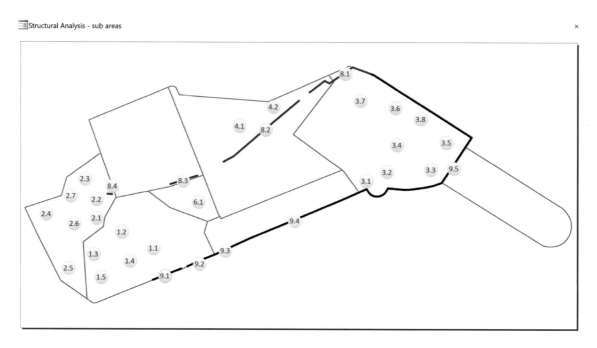

Figure 2. Extract from the GPR metadata form showing easy access to images and animations.

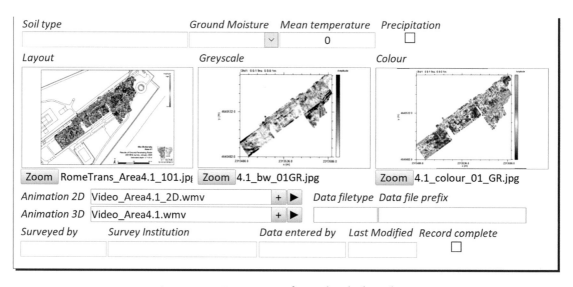

Figure 3. Navigation map for each sub-discipline.

This database not only stores the metadata relating to individual records of, for example, geophysics but also acts as a repository and filing system for all the images and animations produced as shown in Figure 3. Links to the original datasets provide a simple means of relating raw data to the finished product and it is this process, I would argue, that makes the use of a comprehensive set of metadata more than simply creating a repository of dry and obscure data. Providing a tool that enables many additional questions to be asked of the data, provides added value to the research of users of the database and as a consequence encourages them to record this additional data. In the case of archival research, this could simply be a means of grouping data sets of different quality, by attaching a confidence level to the data based on the archival source and/or age of the data. In the case of geophysics, additional data relating to weather patterns before and during survey may have a direct bearing on the quality of data, for example the amount of attenuation of signal experience during a survey GPR survey could be easily demonstrated by the extraction of distinct datasets sharing similar attributes.

The added value in making the extraction of collated data as simple as possible comes from the programming of the query methodology of constantly asked questions within the database. This enables consistency of output for comparison throughout the life of the project. The modular structure of the database also enables the addition of extra elements without affecting the integrity of the database structure. It also allows for the identification and removal of redundant data classes prior to creation of the final archive.

Often detailed metadata is recorded to satisfy the requirements of repositories such as the Archaeological Data Service. Such repositories have specific requirements relating to data type and format and it is only by collecting the metadata with the strictures in mind that the production of the final archivable can be seen as a straightforward task rather than a torturous chore. The concept that by recording metadata in an easily accessible format and integrating it within the visual spatial data generated by a GIS, it is possible to not only produce something that provides immediately tangible results but also produces the means by which a thorough exploration of data integrity can be easily achieved. The grading of data between that which can be deemed empirically precise and that which could at best be described as 'fuzzy' plays an important role in the successful integration of data from a wide-ranging set of disciplines. It is hoped that this paper will show that by making the process of accessing and cross-referencing the metadata from these disparate sources for specialist and non-specialist alike it is possible to holistically substantiate arguments regarding the relative validity and value of data sources within the project.

Acknowledgement

'Rome Transformed' has received funding from the European Research Council (ERC) under the European Union's Horizon 2020 research and innovation programme (grant agreement no. 835271).